AF580359

AFRO-ARABIAN GEOLOGY

A kinematic view

AFRO-ARABIAN GEOLOGY

A kinematic view

Robert Bowen
Ulrich Jux

Institute of Geology
University of Cologne
Federal Republic of Germany

London New York
CHAPMAN AND HALL

First published in 1987 by
Chapman and Hall Ltd
11 New Fetter Lane, London EC4P 4EE
Published in the USA by
Chapman and Hall
29 West 35th Street, New York, NY 10001

Printed in Great Britain at the University Press, Cambridge

ISBN 0 412 29700 0

British Library Cataloguing in Publication Data

Bowen, Robert
Afro-Arabian geology : a kinematic view.
1. Geodynamics 2. Geology—Africa, Northeast 3. Geology—Near East
I. Title II. Jux, Ulrich
551 QE501

ISBN 0-412-29700-0

Library of Congress Cataloging in Publication Data

Bowen, Robert.
Afro-Arabian geology.
Bibliography: p.
Includes index.
1. Geology—Africa, Northeast. 2. Geology—Arabian Peninsula. I. Jux, Ulrich. II. Title.
QE339.N66B69 1987 556.1 87-306

ISBN 0-412-29700-0

"This country is a palimpsest in which the Bible is written over Herodotus and the Koran over that".

Lady Duff Gordon, Letters from Egypt, 1865

CONTENTS

LIST OF FIGURES

PREFACE

"Speak to the Earth and it shall teach thee".
Job, 12, 8.

In 1984, one of us (R.B.) organized an international conference at Whittier College California on geological problems in Libya, Egypt, Sudan, the Levant and Arabia. This was, to some extent, a retrospective of SIR-A, the "Columbia" space shuttle experiment in 1982, and USGS personnel involved with this attended as did other American, German and Egyptian geologists. It was decided to publish the results, but this was not feasible, hence the most significant are included in this work which was completed jointly with one of the invited participants (U.J.). Preliminary reports on the meeting appeared in 1984 (1,2).

The book is a kinematic survey of the geologic evolution of northeastern Africa and the Arabian Peninsula, thus a dynamic investigation of the process of development of the Arabian-Nubian Plate and its separation into two shields with the concomitant sea-floor spreading creation of the Red Sea and subduction of lithosphere along the Zagros trench. The approach is to cover the most interesting and exciting aspects of the Phanerozoic event such as the Ordovician and Permo-Carboniferous glaciations, the recent ice age impact, the marine incursions and retreats coupled with the Mediterranean Miocene desiccation, the rise of the Nile, the Garden of Eden in the Fayum and the spoor of Man out of Olduvai. In addition, recent advances in satellite imaging radar, with the associated revealing of the radar rivers of the Western Desert, and the unravelling of the Libyan Desert glass mystery are examined.

On spelling, US practice is normally followed, but "terrain" is retained in place of the obsolescent "terrane". Arabic words are given their simplest English renderings. Historic dates may be succeeded by B.C. or A.D., prehistoric and earlier by either B.P. or Ma. As all radioisotope values entail uncertainties, the use of "plus", "minus", "circa", "approximately" and "about" is generally avoided. As regards radiocarbon "ages", it seems that the third millenium B.C. was much more radioactive than today, pace Chernobyl, so that 800 years need to be added to a value from this time interval to correct it. It is interesting that, despite high becquerel levels, Man appears to have prospered and indeed the first towns were built then!

An attempt has been made to make a story out of the multitudinous happenings of the vast eon and more with which the book is concerned. Many texts are written in a tedious style, but it is not necessary to use a sterile technique even if an arid region is being described! To enliven matters, odd historic information about the Dynastic period and the ancient Egyptian myths and language, famous figures of antiquity and other appropriate comments have been inserted at suitable points.

On strictly geologic opinions, these are so divided among earth scientists that it is tempting to say, with Terence, "Quot homines tot sententiae". Consequently, the intricacies of the disputes on the stratigraphy and paleontology have been merely mentioned, the names of authors being confined to reference lists at the ends of chapters and allusion to them consisting of a simple number terminating appropriate paragraphs. Mostly, they are from the 1980s, this being a requirement of space restriction. Also, there is a huge fund of earlier data of variable utility already available, hence it would be jejune to quote from them. Authors are usually mentioned only once. This is justifiable because the region was predominantly desert through time, hence the minutiae of its fate are really just variations on a theme.

As regards the index, this is geographic and lists practically all the geologically important places mentioned in the text. However, it excludes subjects of complete chapters such as the Red Sea (IV-9) or the Nile (IV-10) as well as localities peripheral to the region and not considered especially relevant. The aim was to produce a readable, informative, stimulating and reasonably complete treatment of the subject and it is hoped that this book is it!

REFERENCES

1. BOWEN, R., 1984. Geological problems of north-east Africa. Nature, 311, 5982, 108.
2. BOWEN, R., 1984. Meeting reviews African geology. Geotimes, 29, 9, 13-14.

Part I

BASEMENT DYNAMICS

Molten rock or magma is generated deep in the crust and at the upper levels of the mantle of the Earth, white-hot temperatures existing to within 200 km of the planetary surface.
About 70 km down, conditions allow a small quantity of rock to melt, which reduces its strength and permits movement below a relatively rigid cover to take place. This comprises a jigsaw of mobile plates bounded today by mid-oceanic ridges and trenches along the margins of the oceans. In the former, sea-floor spreading creates new crust, while in the latter, crust is destroyed by a process of ingestion called subduction. This pattern is clearly revealed by earthquake, volcano and hot spring distributions around the world.
The same situation most probably obtained through earth history and the early four-fifths of this is discussed in this section with especial reference to the Precambrian basement. This is a crystalline complex underlying younger rocks in the region. In it, granite massifs and eruptives may have resulted from the interaction of subsurface hot materials along subduction zones. It records the development of a colossal, continent-sized block or craton which later became the Arabian-Nubian Shield. There is also evidence of a pair of mountain-building movements. The later of these produced the island-arc, ocean-basin structure of the shield. Associated with this are volcanic and sedimentary accumulations which embrace a widespread occurrence of five hundred million year ages as determined by the potassium-argon dating method. The basement testifies to these echoing a major episode in geologic time, the Pan-African Event of half an eon ago.

Chapter I-1

PRECAMBRIAN OROGENIES

"And thou seest the hills thou deemest solid flying with the flight of clouds".
Surah 27, "The Ant", from the Koran.

THE ASSEMBLING OF AFRO-ARABIA

All Egyptian cosmogenies involve the idea that the universe commenced with a primordial ocean called Nun which was compared to floodwaters and may have been suggested by those of the Nile. Nun filled the universe and comprised a cosmic egg containing stagnant water. It was believed that a primeval hill emerged from Nun and the first pyramids are probably symbolic of this. Could this be an intuitive premonition of Panthalassa and Pangea? The Egyptians imagined that the land would one day again be swallowed up by the primordial waters.
The Afro-Arabian Rift System is superimposed on basement of various types and ages. It includes the Late Precambrian fold belt of the Arabian-Nubian Shield, the Mozambiquian gneisses, the Ubendian, Kibaran and Boganda-Toro mobile belt and the Archean Basement of Uganda. The initial Late Cretaceous dilation in the Afro-Arabian Rift System occurred in the Red Sea, but the rift faults are of ancient, pre-rift origin. The Red Sea possibly opened along a major lineament, the extension of the Marda Fault of Ethiopia and Somalia. The opening of the Gulf of Aden probably was predetermined by a meridional Pre-cambrian lineament which continues west to produce a sharp displacement of the Ethiopian rift at the latitude of Addis Ababa (1,2).
Country rocks in the Red Sea region of the Sudan through which batholithic granitoids intruded may be divided into the Cratonic Gneiss Group and the Metavolcanic-Metasedimentary Complex. The Group is

a part of the Mozambique Belt which extends from East Africa to Sinai and Saudi Arabia. It is best exposed in southern Ethiopia where erosion has stripped off overlying sequences and is probably deeply buried in Saudi Arabia. The Metavolcanic-Metasedimentary Complex is the most widespread unit in the region, the rocks occurring as roof pendants over underlying batholithic granite in belts running northeastward. In Egypt, the rocks are tentatively dated at 1,300 Ma to 1,150 Ma (3).

The mountainous massifs of Gebel Uweinat and Archenu constitute major topographic landmarks defining the threefold junction between Egypt, Libya and Chad. Gneisses and crystalline schists are intruded by granite here. Both plutonic and volcanic rocks outcrop in southwest Egypt ranging from coarse-grained granite at Abu Bayan to basalts at Gebel Uweinat (4).

Two major orogenies have been identified in the basement. The first was in the Middle Precambrian after the sedimentation of pelites and marls now represented by the gneisses of Migif-Hafafit. Emplacement of granite and volcanic rocks have not yet been identified in outcrops in this area, but are found as pebbles in the next sedimentary group. The second attained its maximum in the Late Precambrian after the deposition of impure sandstones and clays in an elongate basin. Ultrabasics were intruded and volcanic eruptions took place. The second orogeny is contemporaneous with the emplacement of gabbros and diorites followed by gray granites. These are referred to as old granitoids and are exposed to the south of Gilf Kebir. Young granitoids are found east of Bir Sahara and there are some dikes and veins cross-cutting a granite. Quartz trachyte and syenite plugs intruded both types of basement rock and also later Paleozoic sandstones, implying a Jurassic to Cretaceous age of foreland platform volcanism. During the Middle Tertiary to the Quaternary, small basic extrusions were widespread in the region of the Gilf Kebir and Gebel Uweinat. They form dikes, cones and sheets of small lateral extent.

Much of northeastern Africa and Arabia is composed of a Precambrian island-arc, ocean-basin complex less than one billion years old. Original oceanic substrate is recognizable as a mafic and ultramafic sequence comprising the oldest units of the complex in the central part of the Eastern Desert of Egypt. This is inferred from elemental concentrations in minimally metamorphosed component rocks. In ultramafics and submarine lavas, contents of chromium, nickel, titanium and the rare earth elements parallel those occurring in contemporary oceanic crust.

The oceanic substrate is covered by thick calc-alkaline volcanics and related volcanogenic meta-sediments. The andesites resemble those of existing Circum-Pacific island-arcs, but contain larger quantities of chromium and nickel. Occasionally,

Figure 1.1: Present rift systems in Africa, the divided Arabian-Nubian Craton (crosses) and its peripheral ophiolite belt (black). Paleogene alkali-basalts are shown by crossed lines.

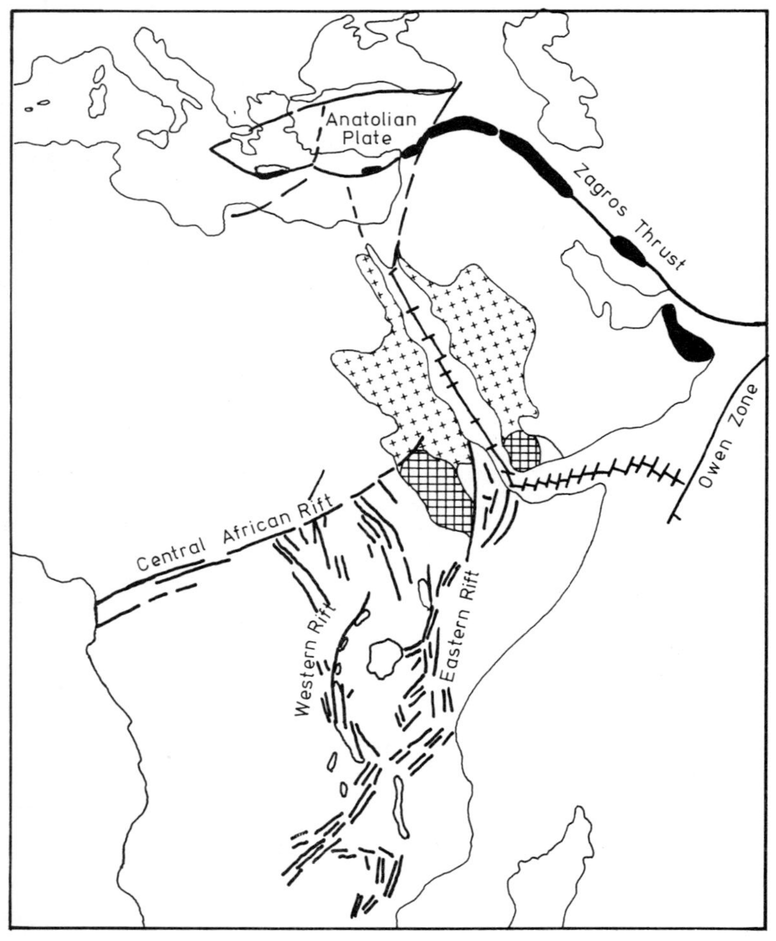

metasedimentary cobble beds in the Eastern Desert of Egypt contain granitic and quartzitic clasts. They derive from the old Proterozoic and Archean forelands, presumably those exposed west of the Nile, for instance at Gebel Uweinat. Stratiform basaltic komatiitic sills are up to 1 km in thickness and, with associated gabbros, are intruded into the metasediments.

The ocean-arc complex of eastern Egypt is both engulfed and surrounded by granitic plutons. They range from syntectonic quartz diorites to a post-tectonic intrusion of large incompatible lithophilic (LIL)-granites emplaced 570 to 550 Ma ago. The acronym means large ion lithophile-rich granites s.s. with 70 to 75% silica, 4 to 5 % potassium oxide and a potassium oxide-sodium oxide ratio near to unity (5).

Radioisotope studies indicate that the Egyptian portion of the ocean-arc complex could have developed, been telescoped and intruded by increasingly more voluminous and fractionated granites between 850 and 550 Ma. In Saudi Arabia, some isotopic dates imply an earlier commencement of these processes. Both the volcanic rocks of the complex and the intrusive granites have low initial strontium isotope ratios of less than 0.704, suggesting a mantle origin. There is no evidence of any older sialic roots under the complex (Figure 1.1).

The Eastern Desert of Egypt is mostly made up of the Late Precambrian Nubian part of the now split, but then united Arabian-Nubian Shield marked by the widespread occurrence of granitic plutons. Their intrusion accompanied the Pan-African event some 650 to 550 Ma ago, discussed in detail in the next chapter. However, in Egypt, their time span may be more limited, of the order of 590 to 570 Ma. Zircon ages from a quartz diorite pluton perhaps pre-dating Pan-African plutonism or representing its initial magmatic stage together with ages from granitic cobbles and a potassium feldspar-rich arkose gave interesting values. The first yielded a concordant date of 770 Ma and was obtained from exceptionally fresh material. In the second, the ages were highly discordant with an upper intercept age for one of 1,120 Ma. Using the episodic lead loss model, a lower intercept age of 580 Ma was obtained and relates to the widespread Pan-African tectonothermal event. As regards the arkose cobbles, the lead isotope ratios indicate a minimum age for the original igneous source terrain of 1.8 eons. The two data points define a chord on a concordia diagram with upper and lower intercepts of 2.3 and 0.69 eons, respectively. The arkosic cobbles have a wide range of ages and no obvious source within the Egyptian basement complex. Probably, they were derived from adjacent continental areas and were deposited in an evolving arc-ocean basin complex (6).

A CRATONIC INCHOATION

The granites in the southern part of the Egyptian Eastern Desert belong to Late Proterozoic syn-orogenic, late- and post-orogenic cycles. A highly differentiated magma source is inferred from the wide range of chemical composition found in them (7).

Geophysical investigations in northern Egypt show that two episodes of folding affected the basement, an older one giving latitudinal folds intersected by a younger set trending ENE. A similarly directed synclinal structure, the Abu Gharadig Basin, cuts through the area and is bounded by two anticlinal uplifts of which the northern is associated with the Qattara Depression and the southern with the Cairo-Bahariya uplift. The deeper Northern Basin is located north of the Qattara Depression. The depth to basement under it varies between 5 and 6.5 km, compared with 3.5 to 4.5 km under the Abu Gharadig Basin (8).

The structures of the Syrian Arc, discussed below in chapter III-2, appear to be intersected by younger northwest and northeast directed faults. They are probably associated with large vertical as well as horizontal displacements. Two northwesterly faults form a graben, namely the Matruh Basin, with a great thickness of sediments. The Nile Valley is structurally controlled by a set of faults roughly parallel to it and bounding the Nile Basin on the west. Probably tectonic uplift is significant in the origin of major topographic depressions such as Qattara and the Bahariya Oasis.

The extent to which older continental crust was involved in creating part of what is now the Arabian Shield is still uncertain. Most of it developed from oceanic crustal material, probably derived from an island-arc sequence during the time interval 900 to 600 Ma. Radioisotopic measurements from a fine-grained granodiorite in the Gebel Khida region show that continental crust of Early Proterozoic age of about 1,630 Ma exists at that locality. Further lead isotope data indicate an earlier, possibly Archean, source for the lower Proterozoic. Although emplacement occurred 1,630 Ma ago, it may have been followed by severe deformation or possibly even Pan-African remobilization at about 660 Ma. An alternative interpretation is that the intrusion took place at 660 Ma and derived from 1,630 Ma old basement rocks by plutonism resulting from tectonic activity in the region. In addition, the lead isotopes demonstrate that the 1,630 Ma crustal rocks inherited material from an older, probably Archean, crust at the time of their formation. Then, the inclusion of mantle material considerably modified the Rb-Sr and Sm-Nd systems so that they now give similar, or only slightly older, apparent ages of 1,800 to 1,600 Ma (9).

In western Saudi Arabia, there are two northeasterly directed Precambrian ensimatic island-arc systems, namely the younger Hijaz and the older Asir, each showing lithotectonic provinces conforming to those occurring in modern examples of plate convergence. The former, 800-700 Ma, comprises a thick fore-arc sequence of turbidites overlying an accretionary prism of metamorphosed and fragmented volcanic and sedimentary rocks together with ophiolites. The latter, less than 900-700 Ma old, are exposed at a profounder erosional level where there are only remnant arc and fore-arc deposits. A frontal-arc assemblage of numerous intrusions of diorites and high-grade metamorphics is interpreted as a high temperature and low pressure belt identifying the Asir arc system. There is also a mafic volcanic assemblage in the back-arc area which seems to delineate an inter-arc basin located between the Taif frontal-arc and a lithologically similar remnant arc west of Bisah (Figure 1.2).

Probably, the Hijaz island-arc collided with the Asir Arc-system along a southeast dipping subduction zone recognized as the Bir Umq-Port Sudan suture approximately 700 Ma ago. During this time, epirogenic movements created an intra-massif basin in the Hijaz Arc-system and an accretionary basin in the Asir Arc-system. These were later filled by volcanic and molasse deposits of the Hadiyah and Fatima Groups, respectively. Convergence proceeding between 700 and 675 Ma ago caused the development of the northeasterly running Samran fold belt. This collision event marks the end of an episode of southeast convergence interpreted as the first phase of the progressive eastward evolution of the

Arabian-Nubian Shield.

Figure 1.2: Palinspastic map of the Arabian-Nubian Craton indicating Precambrian terrains and suture zones. Sedimentary rocks are dotted, volcanics more coarsely dotted, ophiolites black and gneissic areas crossed.

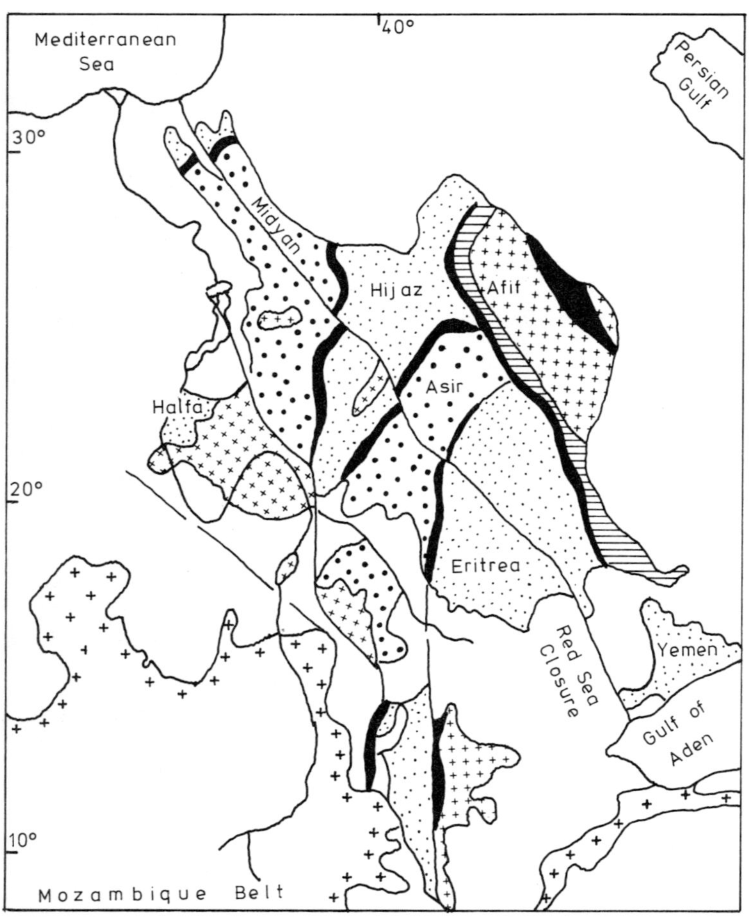

Late Precambrian carbonate rocks in the Gebel Rokhan area southeast of Medina include pseudomorphs after gypsum and nodules of silica. This indicates that evaporite minerals were originally incorporated in the sediment. These probably document the oldest sabkha conditions known. They must have prevailed during the formation of the magnesite deposit found in the upper carbonates. In the same locality, chitinozoan-like microfossils were discovered in a shallow water dolostone. Its age is inferred to be 600 Ma (10,11).

The southern part of the existing Arabian Shield comprises mostly metamorphosed sedimentary and pyroclastic rocks with minor basic and rare acid volcanics. There are five interbedded suites. The first includes minor basic lavas and pyroclastics probably deriving from an immature island-arc environment and the second contains voluminous mafic volcanoclastics originating in that arc. The third is made up of rare acid lavas and pyroclastics while the fourth consists of voluminous quartzo-feldspatic arenites and siltstones probably arising from a continental landmass. Finally, there are minor cherts, quartzites, jaspilites and marbles. Probably, the entire volcanic-sedimentary succession deposited in a marine basin supplied by detritus from an immature island-arc complex on one side and a continental area on the other (12).

Magnetic trend patterns can be used to define magnetic provinces which reflect tectonic terrains. The rocks of the shield include Precambrian to Recent and the oldest are the gneisses. Substantial regions are occupied by older granites which have been dated as 1,000 Ma on the Arabian side and 1,000 Ma to 900 Ma for corresponding rocks on the Nubian side. This points to a plutonic episode which may correlate with or be slightly younger than the African Kibaran Orogeny of 1,300 Ma to 900 Ma ago (13).

The younger plutonic rocks possess syntectonic components related to the Hijaz Orogeny as well as post-tectonic ones related to the Najd Orogeny. Dates of these are believed to be around 800 Ma to 650 Ma on the Arabian side and 600 Ma to 500 Ma on the Nubian side. This orogenic turbulence appears to synchronize with the similar, multi-episodic, East African or Mozambiquian Orogeny, which extended northward from this region into Arabia (14 ; Figures 1.3, 1.4).

Mesozoic and Tertiary rocks occur along the Red Sea coast, the former recording practically no evidence of tectonic activity. This probably reflects the stationary position of the continent during that time interval. However, in the Tertiary, Red Sea rifting may have been initiated by updoming of the adjacent parts of the Arabian-Nubian Shield.

Magnetic analyses showed major preferred orientations in both Arabia and Nubia implying a single origin for the shield. A latitudinal trend is the result of a northern horizontal compressive force

Figure 1.3: Granitic mountains near the convent of St Catharine, Sinai.

Figure 1.4: Weathered blocks of granite showing exfoliation located north of Gebel Uweinat in the Libyan Desert.

with a major vertical release. Another one, ENE, is believed to be younger than the latter and to relate to a northwesterly shift of the same compressive force. This accounts for the situation in the north. However, the same trend to the south may result from a compressive force arising from the opening of the axial trough in the southern Red Sea.

A WNW direction is considered to be associated with the force of couple which affected the Arabian-Nubian Shield. The northwesterly structure is interpreted as a major shear fracture resulting either from a northern compressive force with major release in an east to west horizontal direction or from a vertical compressive force with major release in a meridional direction. This trend was recorded on both sides of the Red Sea and diminishes away from it.

The Tarr complex forms the southernmost unit of the Late Proterozoic Kid Group in the southeastern Sinai Peninsula and, north of it, there is a volcano-sedimentary pile showing features of an island-arc succession. However, the northward terminating unit resembles a marginal basin ophiolite. The culmination of volcanic events produced northeast directed, fault-controlled highlands. The complex is composed of coarse and fine volcanic debris, was deposited along the southern and southeastern flanks of the arc edifice in newly formed proximal basins and the chaotic state of the rocks attest to a fault scarp mélange olistostromal system formed during regional gravitational slumping of the volcano-sedimentary pile. This phase designates a major change from arc volcanism to arc degradation. Sediments were shed southward to create submarine fans, clastic wedges and tidal flats in tectonically bound troughs. The sedimentologic, tectonic and metamorphic history of the Tarr complex is connected with subduction and collision. Similar rocks outcrop in Jordan and include major intrusions of plutonites and subsequent dikes. Late Proterozoic sediments unconformably overlie them and were again intruded by hypabyssal rocks. This whole Basement Complex belongs to the northern margin of the African Craton which consolidated rather late.

In the region of Wadi El-Sheikh, southwestern Sinai, older and younger granites can be recognized. The latter were intruded during two phases. Zircon crystals become progressively more uniform in shape in the younger granitic rocks and can be used as a petrogenetic guide in elucidating the origin of the intrusions (15,16).

During the Late Precambrian, earlier continental crust was destroyed and various types of rifting, including Red Sea-type, probably occurred. Although only one single vast landmass, Pangea, initially existed, the first differences in the mobile zone distribution patterns in Gondwana and Laurasia became evident in the Early Riphean. It is alleged that this may be seen from greenstone belts and other linear intracontinental structures. Some earth scientists believe that separation of the daughter supercontinents had already taken place during the Late Proterozoic, the fragmentation of Laurasia and the opening of the Iapetus and the Paleo-Asian Ocean coinciding in time with the Pan-African Orogeny and its equivalents in Gondwana. This hypothesis is contrary to the generally accepted view that Gondwana split off from Laurasia a mere 200 Ma ago (17).

REFERENCES

1. ENGEL, A.E.J., DIXON, T.H. and STERN, R.J., 1980. Late Precambrian Evolution of Afro-Arabian crust from ocean arc to craton. GSA, Bull., 91, 699-706.

2. KAZIM, V., 1979. Relationship between rifts and Precambrian basement in East Africa. Ann. Geol. Sur. Egypt, 9, 54-60.

3. JAR EN NABI, M., HAILU, T., 1979. Contributions to the stratigraphy of the basement rocks in the Red Sea region: Sudan. Ann. Geol. Surv. Egypt, 9, 45-53.

4. DARDIR, A.A., 1982. Basement rocks of the Gilf-Uweinat area. In: Desert Landforms of southwest Egypt: A basis for comparison with Mars, ed. El-Baz, F., Maxwell, T.A., NASA, CR-3611, 67-78.

5. DIXON, T.H., MEGUID, A.A.A. and GILLESPIE, J.G., 1979. Age, chemical and isotopic characteristics of some pre-Pan African rocks in the Egyptian Shield. Ann. Geol. Surv. Egypt, 9, 591-610.

6. HEIKAL, M.A., SALEM, A.-K.A. and El-SHESHTAWI, Y. A..1985. Zircon in granitoids from Sinai, Egypt and its genetic significance. Szeged Acta Min.-Pet., 27, 117-130.

7. SALEM, A.K.A., KABESH, M.L. and GHAZALY, M.Kh.,

1985. Petrochemistry, petrogenesis and classification of Um Huqab, Garf and El-Mueilha granitic masses, Southeastern Desert Egypt. Szeged Acta Min.-Pet., 27, 17-31.

8. MESHREF, W.M., REFAI, E.M., SADEK, H.S., ABDEL-BAKI, S.H., El-SIRAFE, A.M.H., El-KATTAN, E.M. I., El-MELIEGY, M.A.M. and the late El-SHEIKH, M.N., 1980. Structural geophysical interpretation of basement rocks of the north Western Desert of Egypt. Ann. Geol. Surv. Egypt, 10, 123-937.

9. STACEY, J.S., HEDGE, C.E., 1984. Geochronologic and isotopic evidence for Early Proterozoic crust in the eastern Arabian Shield. Geology, 12, 310-313.

10. BOKHARI, M.M., BINDA, P.L. and SCHELLEKENS, J.H., 1981. Evidence of former evaporites in Late Precambrian carbonate rocks of Jabal Rokhan, Saudi Arabia. Jeddah Bull. Fac. Earth Sci., 4, 149-158.

11. BINDA, P.L., BOKHARI, M.M., 1980. Chitinozoan-like microfossils in a Late Precambrian dolostone from Saudi Arabia. Geology, 8, 70-71.

12. RAMSAY, C.R., BASAHEL, A.N. and JACKSON, N.J., 1981. Petrography, geochemistry and origin of the volcano-sedimentary succession between Jabal Ibrahim and Al-Aqiq, Saudi Arabia. Jeddah Bull. Fac. Earth Sci., 4, 1-24.

13. MESHREF, W.M., ABDEL BAKI, S.H., ABDEL HADY, H.M. and SOLIMAN, S.A., 1980. Magnetic trend analysis in northern part of the Arabian Nubian Shield and its tectonic implications. Ann. Geol. Surv. Egypt, 10, 939-953.

14. CAMP, V.E., 1984. Island arcs and their role in the evolution of the western Arabian Shield. GSA, Bull., 95, 913-921.

15. SHIMRON, A.,E., 1981. Geology and tectonics of the Tarr complex, SE Sinai Peninsula. Geol. Surv. Israel, Current Research, 2, 73-82.

16. SHIMRON, A.E., 1984. Evolution of the Kid Group, southeast Sinai Peninsula: Thrusts, mélanges, and implications for accretionary tectonics

during the Late Proterozoic of the Arabian-Nubian Shield. Geology, 12, 242-247.

17. BOZHKO, N.A., 1986. The evolution of the mobile zones of Gondwana and Laurasia in the Late Precambrian. Tectonophysics, 126, 125-135.

Chapter I-2

PAN-AFRICAN EVENT

"Arabia is like a huge rough-hewn paving-stone which has been slightly tilted so as to dip towards the south-east".
Arnold Toynbee, East to West, 1958.

GENESIS OF A THALASSOCRATON

Previously, Precambrian cratonization in Afro-Arabia was described, a process involving the formation of crust. The ancient Egyptians saw the matter in a somewhat different light. In the Book of the Dead, Shu, God of the atmosphere, is shown raising his daughter Nu, Goddess of the sky away from Geb, God of the Earth. Thus, the planet differentiated and became independent from the heavens (Figure 2.1).

Figure 2.1: The creation of the Earth in Egyptian mythology.

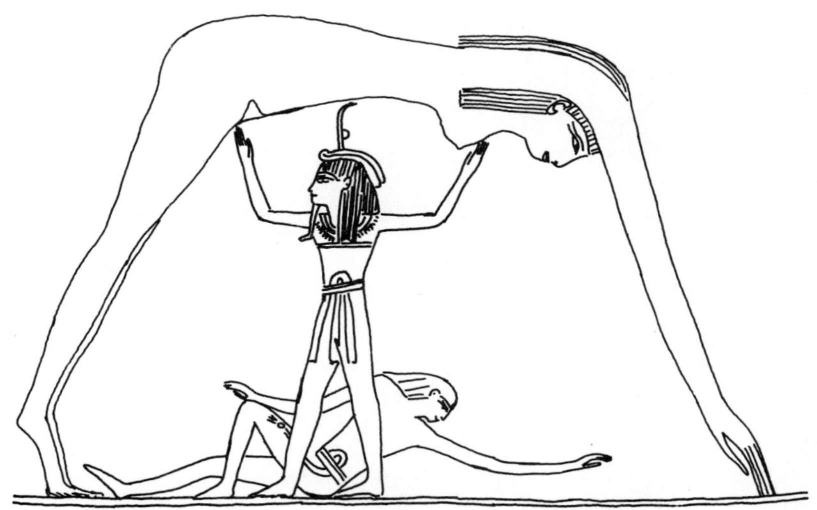

The term "Pan-African" describes a major geologic event. It produced the extensive occurrence of 500 Ma K-Ar ages in Africa. Later broadening of the usage requires a redefinition, but this does not affect its occurrence throughout Africa and Arabia. Similar revolutions took place in other parts of Gondwana, for instance the Brazilian Event 550 Ma ago. The Pan-African superstructure includes volcano-sedimentary and magmatic rocks of Late Proterozoic age. They comprise a western foreland underneath which may continue the Mozambique Belt. Its lithologies, structures, metamorphism, isotope ages and mineral deposits are traceable into Ethiopia and there is some evidence that they extend into the Sudan, Egypt and the Arabian Peninsula.
Although Pan-African ages for granitic rocks of northeastern Africa are often attributed to remobilization of Archean sialic crust, such an origin is incompatible with the limited strontium isotopic data available. The Pan-African belt is part of the Pangeaic orogenic system characterized by emplacement of plutonic igneous rocks with calc-alkaline to alkali-rich compositions. Plutons of this type having low initial strontium isotope ratios are believed to develop in magmas derived from the upper mantle or lower crust. Six such plutons, which represent various types of post-tectonic alkali granite from the Red Sea Hills, gave Rb-Sr whole rock isochron ages of 595 to 568 Ma. Five of these Younger Granites had initial strontium isotope ratios of 0.7016 to 0.7025 and the other gave an initial ratio of 0.7061. Such low ratios could not have been produced by the mobilizing of Archean sialic crust 600 to 550 Ma ago, but could have been produced if the granites came from a source with a low Rb-Sr ratio like the upper mantle or lower crust (1).
By far the best known of these plutonics is the pink and gray ornamental granite capped in places by the Nubian sandstones outcropping where the First Cataract of the Nile at Aswan used to be visible. In earlier times, Pliny referred to it as "syenite" after the ancient name of this city. Granitization here must have succeeded an intense folding and shearing of country rock which was simultaneously injected by acidic fluids stemming from the anatexis of deeper sediments. After this and accompanying diastrophism and metamorphism, there was a period of quiet only interrupted by dike and sheet intrusions associated with a younger granite which was rather homogeneous and aplitic. This fine-grained granite

resulted from the consolidation of intrusions from an anatectic magma linked with late isostatic uplift and orogeny. It is noteworthy that the older pink and gray granites, the former being sometimes referred to as "rose" in color, were often used in antiquity for the sculpturing of gigantic monoliths such as the obelisks which now find their homes in major cities such as Rome, Paris, London and Istanbul.

The only known ages of over 2,000 Ma in the region are for samples from Gebel Uweinat which is shared by Libya, Egypt and Sudan. Two gneisses gave Rb-Sr ages of 2,060 Ma from a feldspar and 2,507 Ma on a whole-rock sample. Finally, a Rb-Sr whole-rock isochron age of 2,620 Ma on charnockitic gneisses from Gebel Uweinat has been proposed as a minimum age for these rocks. In fact, Gebel Uweinat may be just a cratonic fragment rather than an exposed portion of a much larger Archean shield (2).

The Ben Ghnema batholith comprises a series of plutons intruded along a meridional axis in the northwest corner of the Libyan part of the Tibesti Massif. It separates metasediments of platform facies on the west from mafic deposits in offshore facies on the east. This calc-alkalic batholith, which is practically identical lithologically and chemically to the Sierra Nevada batholith of California, has a Pan-African age with an initial strontium isotope ratio of 0.705 to 0.706 (3).

In the batholith, there is a rough gradation from tonalite and granodioritic rock on the east to granite on the west from which it may be inferred that the feature formed during the subduction of oceanic crust westward under a continental craton. The eastern periphery of this, the East Saharan craton, a block of very ancient crust, probably was located in the northwestern Tibesti Massif in Pan-African time. These considerations strongly suggest that this margin was then the border between stable continent to the west and ocean basin to the east.

In western Saudi Arabia, the as yet unsplit Arabian-Nubian Shield comprised widespread, mainly basaltic, metavolcanic and pyroclastic sequences. They are associated with serpentinites and metasediments intruded by batholithic masses of diorites, tonalites, trondhjemites and granodiorites. These sequences are pierced, particularly in the north, by late tectonic calc-alkalic granite-gabbro and alkalic-granite complexes. These rocks formed under island-arc and associated oceanic conditions. This is evidenced by the fact that many of the accompanying mafic to ultramafic masses are ophiolitic fragments. One such is the Hulayfah-Hamdah Ophiolite Belt which is divided into northern and southern parts by the offset of a prominent Najd fault. The belt cuts the present Arabian Shield from north to south and is marked by magnetic anomalies with amplitudes of 2 to 5 X 10^{-3} oersted. Among other ophiolite complexes in the region are the Bir Amq, also cut by a Najd fault, and the Gebel el Wask, this latter probably comprising part of an ophiolite belt which continues into the Eastern Desert of Egypt. They can be followed along extensive, narrow belts which, in their turn, delineate terrains of contrasting volcanic-sedimentary rock sequences. Cratonization took place by juxtaposition and welding of volcanic island-arcs with associated trench-fill sediments (4,5).

In the Sudan, several regions of island-arc and marginal type volcanic, sedimentary and plutonic activity have been proposed. They are separated by ophiolites. On the west, gneissic and metasediment-ary basement may be of continental Proterozoic crust of 1,550 Ma of age. In southeastern Egypt, fragments of even older continental crust have been identified by detrital zircons and dated at 1,800 to 1,700 Ma of age. Nd-Sm analysis yielded a similar age, but the metamorphic nature of gneisses and schists imply a much younger age of 1,185 Ma near regions of disturbed volcano-sedimentary-ophiolite sequences.

The western side of the existing Red Sea has not been investigated as well as the eastern side. Notwithstanding, there is enough evidence to permit rock units like those in the Arabian Shield to be identified. In this way, the palinspastic reconstruction of the shield can be made, allowing establishment of coeval tectonic events across the Red Sea closure.

THE ARABIAN SHIELD

A number of distinct geological provinces can be determined in the Arabian Shield now. Its easternmost portion is underlain by volcano-sedimentary sequences of probable margin type and also by remobilized, tectonized crust of continental affinity. The isotopic studies show mixing between ensimatic and continental material giving an Early Proterozoic age of 1,630 Ma. The now divided Arabian-Nubian Shield originally comprised three main units. These are gneissic and volcano-sedimentary terrains of island-arc and plain margin types as well as ophiolite assemblages. The ultramafic bands of the latter are interpreted as suture zones between intra-oceanic plates or at the margins of continental-island-arc terrains except in the Eastern Desert of Egypt. There is an allochthonous ophiolite sequence comprising serpentinized peridotites, gabbro complex, sheeted diabase dikes and pillowed basalts exposed at Wadi Ghadir. It contains both terrigenous and oceanic components indicating deposition in an oceanic trench over a subduction zone at a continental margin. Subduction took place from the northeast or east.

There, the Precambrian Basement consolidated as a result of early orogenic movements to form an enormous Iranian platform considered to be an extension of the Arabian Shield. During the Paleozoic, only epeirogenic movements occurred with accompanying deposition of a platform deposit. In Central Iran, there were precursory Alpine movements in the Mesozoic, trends within these following the Precambrian structural plan. The Alpine orogeny proper developed in the Late Cretaceous and Tertiary, except in the rigid, eastern Lut Block. In Saudi Arabia, the Nabitah Mobile Belt is a band of deformation, remobilization and plutonism, defining the boundary between continental gneisses on the east and oceanic volcanic sequences on the west.

Late Precambrian volcanism in the central Eastern Desert represent primitive, ensimatic igneous activity on thin crust less than 20 km thick. The metavolcanics closely resemble modern calc-alkaline sequences and very probably were erupted in an arc-like setting. The chemical composition of the relict minerals implies that they crystallized from sub-alkaline, probably calc-alkaline or tholeiitic, magmas. The clinopyroxenes demonstrate abundances of minor elements which resemble those of modern ocean floor, marginal basin or island-arc lavas.

Five oceanic terrains can be identified. Some are concealed under younger Phanerozoic cover rocks. They are separated by four ophiolite-bearing suture zones. Their mutual margins are roughly 200 km apart and parallel and make up the Late Proterozoic Arabian Shield. In the western shield, three ensimatic island-arc terrains are located, but those two located in the eastern shield are continental in their affinities. The two western sutures join island-arcs, whereas the two eastern sutures together constitute a major collisional orogenic belt.

Figure 2.2: The Fatima Formation exposed in the wadi of the same name near Mecca in Saudi Arabia.

Figure 2.3: Pleistocene valley terraces and schists of the Basement Complex in the Wadi Firan Sinai.

Pan-African event

These five accreted into an Arabian craton between 715 and 630 Ma ago. This fusion was coupled with collision-related intracratonic deformation and magmatism. Molasse formed, accompanied by episodes of intermediate to acidic volcanism and occasional, subsequent intrusions of peralkaline to peraluminous granites from 640 to 570 Ma ago. A major left-lateral wrench fault system developed between 630 and 550 Ma ago, displacing the northern section of the Arabian Shield approximately 250 km to the northwest.

Ophiolite lineaments could be ancient subduction zones although, if so, the direction of their inclination is undecided. Of course, it may be that not all ultramafic assemblages are either ophiolites or subduction-derived. Linear ophiolite belts obliquely intersect continental margins and regionally subtend steep angles. If oceanic terrains represent evolutionary accretion and cratonization, there should be a temporal progression of events across them. There are not enough isotopic age determinations to provide a sequential history either between or within their volcanic belts. In the Arabian Shield, the Midyan one is thought to be the youngest at 700 to 570 Ma. Its three volcano-sedimentary sequences involved island-arc accretion and compressional tectonics up to about 625 Ma and extensional tectonics and rifting later. Radioisotope dating appears to indicate that the Murdama, Shammar and Fatima Groups belong to the Vendian Period. The Jubaylah Group in Saudi Arabia is either close to the Precambrian-Cambrian boundary or Late Cambrian in age. The Hijaz is older at 800 to 700 Ma and the southern Asir containing the oldest arc assemblages is more than 900 to 800 Ma of age (Figure 2.2).

The Omani Huqf Group corresponds to the Fatima and is the oldest known sedimentary sequence on the crystalline basement in that country. It outcrops on a broad regional high on the southeastern edge of the Arabian Peninsula. The sediments are Late Precambrian to Middle Cambrian in age. Within it are clastics and carbonates deposited in shallow marine or fluviatile conditions and terminated with an evaporitic sequence. Evaporites do not occur on the axis of the high, but are thickly developed to the west over much of central and southern Oman where most of the oilfields are located. The Huqf Basin may be part of a series of evaporite basins and intervening carbonate platforms which stretched across Pangea from the Indian subcontinent through

South Yemen, Oman, Saudi Arabia and Iran (6,7,8,9). In the Sudan, the volcano-sedimentary sequences may be oldest in the northern Red Sea Hills, about 720 Ma in the central block, and youngest in their southern region. There could be a threefold divison of the Eastern Desert and northern Red Sea Hills in which the basement becomes younger to the north, ranging from 765 Ma in the south to 540 Ma in the north. Large tholeiitic dike swarms occur in basement rocks in the Red Sea Hills and have northwesterly, latitudinal and meridional trends. Two periods of emplacement are proposed, one at 660 and the other at 616 Ma ago. These dates suggest an association with the northeastern African igneous ring-complex province.

Maximum initial strontium isotope ratios increase sharply after 700 Ma, these mostly occurring in the northeastern Arabian Shield. This trend evidences an older sialic crust in this area. Also, massive sulfide deposits in the Arabian Shield have been invoked in support of the existence of a 2,100 Ma continental crust in the eastern shield. This hypothesis cannot be discounted, but model Rb-Sr ages from suites of high initial strontium ratios suggest that their source region separated from the upper mantle less than 100 Ma before their emplacement (10,11,12).

The recognition of granitic rocks in the eastern part of Arabia with initial strontium ratios over 0.705 could well indicate the presence of much older crust. As no 800 Ma granodiorites with rubidium-strontium ratios of unity have yet been detected in the eastern part of the exposed shield, some do not accept an anatectic origin of granodiorites with this ratio as having been emplaced 100 Ma earlier. Nevertheless, rocks of this age do occur in the southern part of the Arabian Shield, but these have Rb-Sr ratios of less than unity and usually under 0.2. The conclusion is that the presence of granitic rocks in the eastern part of Arabia with initial strontium ratios exceeding 0.705 probably indicates the occurrence of much older crust.

In the southeastern At-Taif, Pan-African granites represent the final episode in the Hijaz tectonic cycle and comprise the youngest and most abundant rock unit there. This magmatic arc is a Late Proterozoic belt of metasediments, calc-alkaline volcanics and ophiolites intruded by granitoid rocks of tonalitic, alkaline and peralkaline composition. It extends from Saudi Arabia and the Eastern Desert of Egypt to the Sudan. Diorite and metamorphic rocks

are mostly enclosed in the granites. The chemical data suggest that the granitic rocks evolved under compression by partial melting of the lower crust. The intermediate rocks may have arisen through the interaction of melts derived from the oceanic crust and mantle with ascending magmas (13).
Amphibolitic and quartzo-feldspathic gneisses were intruded by three synkinematic and one postkinematic groups of calc-alkalic granite, isotopically dated between 595 and 525 Ma. Initial strontium ratios of 0.703 to 0.710 imply an origin by either partial fusion of the mantle or lower crustal anatexis. Field, geochemical and petrographic work revealed emplacement in an island-arc environment. The crystalline basement of Arabia and northeastern Africa results from the final cratonization of island-arcs over a 600 million year period from 1,100 to 500 Ma ago. Thus, Precambrian orogenesis and plutonism lasted at least for a time exceeding by an order of magnitude that taken by such events during the Phanerozoic. The continental gneisses on either side of the Pan-African oceanic mosaic are Early to Middle Proterozoic, but, like the entire region, they have been remobilized, metamorphosed and widely disrupted by the tectonothermal events of the Late Pan-African 600 to 550 Ma ago. Favorable economic mineral domains can be postulated on the basis of regional tectonic structure and certain of the plates contain volcanogenic sulfide deposits, ultramafic-hosted chromite and hydrothermal quartz-gold mineralization. In Tertiary times, the Red Sea opened across the volcano-sedimentary-ophiolite belts almost along the central line between the fringing Arabian and Nubian Shields, remnants of the former craton which by now had separated.
In Israel, the crystalline basement under the central and northern parts of the country forms a distinct high. It plunges toward the Mediterranean, the Dead Sea Rift and a local depression in the north. Probably pre-Late Jurassic tectonic movements occurred and cannot be detected in the surface geology. Variations in thickness of sediments provide additional evidence for a deep-seated structural ridge under the eastern coastal plain as well as two deep step-like zones near the southern and northern boundaries of the uplifted area of Central Israel. Such old tectonic features and their related lithostratigraphic lateral variations could have produced conditions favorable for entrapping hydrocarbons (14).

THE NUBIAN SHIELD

In the Western Desert of Egypt, there were four deformational episodes with related metamorphism and anatexis until cratonization ended in the last stage of the Pan-African crustal upheaval. There may be as many as six separate magmatic episodes up to the Quaternary. This anorogenic type of magmatism is primarily related to a fracture system originating in the Late Precambrian through intraplate block faulting. Periodic reactivation of the older fracture zones through the Phanerozoic produced the different types of plutonic and volcanic rock assemblages (15).

The basement rocks of Egypt can be subdivided into the Meatiq Group (oldest), the Abu Ziran Group and the Hammamat Group (youngest), the last two belonging to the Pan-African orogenic cycle. The Meatiq Group is old crystalline basement outcropping in gneiss domes. They probably evolved in a mainly compressional tectonic environment during the Late Proterozoic and exhibit many of the basic structural and lithological characteristics of Cordilleran metamorphic core complexes. They comprise an anticline with low dipping foliation and unidirectional mineral-slickenside lineation. The core comprises granitic gneiss conformably overlain by a heterogeneous and isoclinally folded mylonitic material which predates the doming event. This grades up into low-grade ophiolitic rocks. The area lies in the foreland fold and thrust belt of a continental margin origin. Ophiolites outcrop along the thrust between the Meatiq infrastructure and the imbricated Abu Ziran nappe. Calc-alkaline magmatism occurred along two upwarps and was associated with gold mineralization (16).

The Abu Ziran Group comprises ophiolites overlain by metasediments, pyroclastics and local intermediate volcanics with island-arc characteristics. The Hammamat Group includes molasse-type clastics and penecoeval Dokhan Volcanics of andesitic to rhyolitic composition, these being equivalent to plutonic, syn- to late-tectonic calc-alkaline granites. The clastics are up to 5 km thick, of Late Pan-African age and exposed in the coastal mountains of the Eastern Desert. There are four lithofacies which are conglomerate, pebbly sandstone, sandstone and siltstone, respectively. The interrelationships suggest alluvial fan-braided stream deposition within several small-sized basins. The interbedding of conglomerate and thick siltstone units indicate the direct interdigitating of fans with playas or lake sediments. Discontinuous siltstone units are interpreted as cut-off channel deposits within braided streams. Debris flow sedimentation is not exhibited. Paleomagnetic directions have been identified at several sites in the Dokhan Volcanics and also from two dike swarms intruding the late orogenic (younger) granites. The reported Rb-Sr ages for the Dokhan Volcanics range from 660 Ma to 603 Ma. The higher coercivity component of these gave a paleopole at 36° S, 17° E, which is not easy to reconcile with the accepted versions of the African polar wander path. The pole of the softer component, however, is at 54° N, 327° E, suggesting a Cambrian age of magnetization.

The dike swarms, which range from 530 Ma to 480 Ma in age, have poles at 87° N, 304° E and at 86° N, 185° E, respectively. It is believed that the magnetization is primary, even though the paleopoles of the dikes are close to the present geometric pole. These data, when compared with others from Africa, imply that the African polar wandering formed a loop in the vicinity of the present Arctic about 600 Ma to 500 Ma ago. Also, it suggests that the Nubian Craton was behaving in a manner consistent with a single African Plate at the time. It may be added that the intensity of natural remanent magnetization was often remarkably high (up to 55 X 10^{-2} emu/cc), but such rock samples were either magnetically unstable or gave anomalous directions of magnetization which could be attributed to remagnetization by lightning (17).

The Abu Zawal area is approximately 150 km northeast of Luxor and is composed of a sequence of igneous and metamorphic rocks of Late Precambrian age. The least radioactivity is associated with basic metavolcanics and the highest values relate to late orogenic plutonites. the granodiorites possess a moderate level of radioactivity. Joint investigations show that the complementary of all forces acting on the area strikes in a northwesterly direction (18).

In the Eastern Desert of Egypt, there are four episodes of gold ore formation which can be recognized, namely, the pre-orogenic, the syn- to late orogenic, the Late Proterozoic to Early Paleozoic and the Mesozoic to Cenozoic ones. As well as the widespread quartz-vein type, there are gold occurrences in sulfides and ferruginous quartzites. Hydrothermal processes in many areas produced potentially auriferous zones from 10 to 100 m thick which extend as long as the quartz veins. The highest gold content recorded derived from primary hydrothermally altered rocks, in this case coarse-grained biotite granite, at Um Samra and yielded up to 24 ppm. Gold mining started during Pharaonic and Roman times and ceased as late as 1958. From 1902 until that date, 6.9 tons of gold, with an average content of between 10 and 30 ppm were extracted.

Figure 2.4: Pharaonic map of the gold-mining area in Wadi Hammamat, Eastern Desert, Egypt.

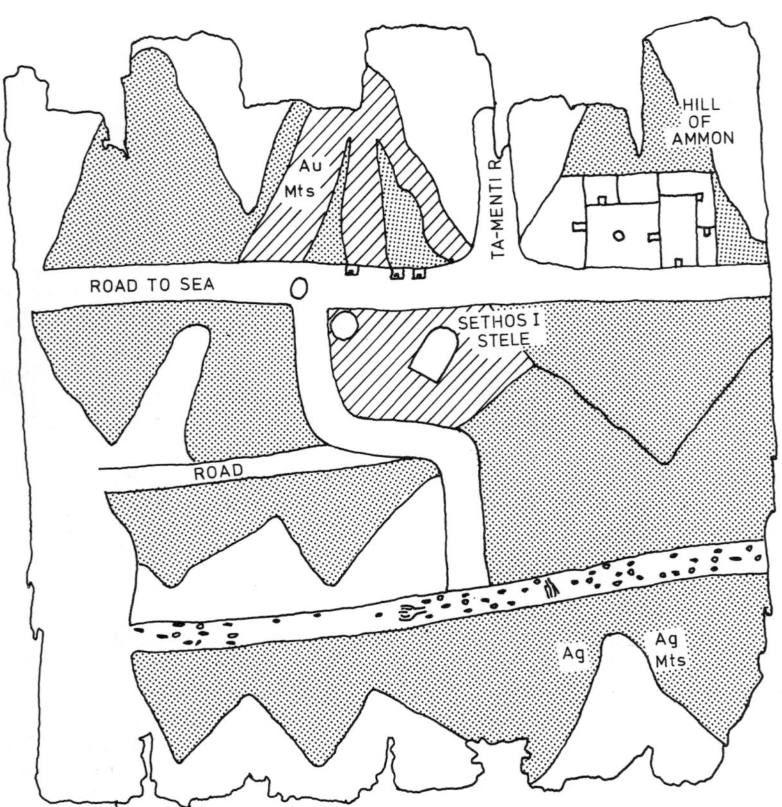

Gold and silver were the only precious metals known from the antique world and, of these two, silver was much rarer and hence more valuable. Egypt possessed the greatest gold reserves at that time and its people were highly capable prospectors and miners. In fact, by the 19th Dynasty, accessible supplies had been exhausted and Sethos I together with his son Rameses II had to initiate mines further away in the Eastern Desert. One of these may well have been in the Wadi Hammamat and is recorded by the Turin Papyrus, the oldest geological map so far found. Unfortunately, this is difficult to prove because there are more than one hundred old mines in the Eastern Desert. Nevertheless, the one in question must be near the bekheny-mountain shown on the papyrus and the nearest locality is Umm Fawakhir in that wadi. On the map, this mountain is colored black and the area around the mine is red, this probably indicating the different rock types exposed. There is an abrupt change from the dark wadi rocks to the pink granites around the mine at Fawakhir. Various objects shown on the papyrus as littering roads may well convey the idea of alluvial material in a trunk wadi.

The Egyptian gold is far from pure and often contains silver, sometimes to the extent of 20% of the whole when a pale amber-colored alloy known as electrum is the result. As well as occurring in the Eastern Desert, it is also recorded from the land of Punt which may have been Somalia. At first it was thought to be a separate metal, but, by the New Kingdom, it was being produced artificially. Curiously, the ancient Egyptians believed that silver was a form of gold and they called it "white gold". It is interesting that practically all the silver from the Old and Middle Kingdoms contains gold, compositions varying from 38 to 9%.

All these valuable metals were used for setting gemstones of generally semi-precious character chosen for color rather than reflective power. By far the most popular jewellery embodied the blood-red carnelian, the blue-green turquoise and the cerulean lapis lazuli. This latter was imported from Afghanistan via Mesopotamia and the other two are found in Egypt. Carnelians can be found with no problem in the Eastern Desert, but turquoise was mined from Early Carboniferous dolomites near presently exploited manganese ores in Sinai. From the basement, there came transparent rock crystal, green amazonite and beryl, purple amethyst, red garnet and jasper in various colors. Hathor, the Goddess of Love, Beauty and Mining, was especially associated with jewellery, her priestesses carrying elaborate menyet necklaces as cult symbols (Figures 2.3, 2.4).

The Precambrian Basement along the northwestern Gulf of Suez is inferred to represent rapid continental crust accretion from 670 to 550 Ma. A variety of considerations such as field and isotopic data demonstrate that the event occurred in a tensional stress setting. The evolution of crust in this region is therefore different from the situation in other parts of the Arabian-Nubian Shield in which the crust seems to have formed by convergent margin accretionary processes and collisional tectonics. The Pan-African granites represent a non-orogenic magmatic pulse not related to subduction zone magmatism.

In the Precambrian around the Red Sea, there are three major metallogenic cycles. The earliest is the Mozambiquian, 2,000 to 550 Ma old, comprising stratiform base metal deposits, metamorphic minerals such as graphite and pegmatites with gemstones, micas and rare metals. The Pan-African volcano-sedimentary event reached its peak 700 Ma ago. It is characterized by banded iron-ores and also by syn- and epigenetic auriferous as well as argentous sulfides associated with an episode felsic island-arc volcanism. The Pan-African produced major rare metal deposits of tantalum and tin as well as some of the gold-quartz veins. These cycles overlap spatially and chronologically which complicates identification.

REFERENCES

1. FULLAGAR, P.D., 1980. Pan-African age granites of northeastern Africa: new or reworked sialic materials? In: The Geology of Libya, 3, ed. SALEM, M.J., BUSREWIL, M.T., 1051-1064, Acad. Press, London, UK.

2. SCHANDELMEIER, H., DARBYSHIRE, F., 1984. Metamorphic and magmatic events in the Uweinat-Bir Safsaf Uplift (Western Desert, Egypt). Geol. Rdsch., 73, 819-831.

3. ROGERS, J.J.W., HODGES, K.V. and GHUMA, M.A., 1980. Trace elements in continental-margin magmatism: Part II. Trace elements in Ben Ghnema batholith and nature of the Precambrian crust in central North Africa: Summary. GSA, Bull., 91, 445-447.

4. VAIL, J.R., 1985. Pan-African (Late Precambrian) tectonic terrains and the reconstruction of the Arabian-Nubian Shield. Geology, 13, 839-842.

5. El-SHARKAWY, M.A., El-BAYOUMI, R.M., 1979. The ophiolites of Wadi Ghadir area Eastern Desert, Egypt. Ann. Geol. Surv. Egypt, 9, 125-135.

6. HARRIS, N.B.W., HAWKESWORTH, C.J. and RIES, A., 1984. Crustal evolution in N.E. and E. Africa: Evidence from model Nd ages. Nature, 309, 773-776.

7. KRÖNER, A., 1985. Ophiolites and the evolution of tectonic boundaries in the Late Proterozoic Arabian-Nubian Shield of northeast Africa and Arabia. Precamb. Res., 27, 277-300.

8. BINDA, P.L., 1981. The Precambrian-Cambrian boundary in the Arabian Shield: A review. Jeddah Bull. Fac. Earth. Sci., 4, 107-120.

9. GORIN, G.E., RACZ, L.G. and WALTER, M.R., 1982. Late Precambrian-Cambrian sediments of Huqf Group, Sultanate of Oman. AAPG, Bull., 66, 2609-2627.

10. DUYVERMANN, H.J., HARRIS, N.B.W., 1982. Late Precambrian evolution of Afro-Arabian crust from ocean arc to craton: Discussion. - Stern, R.J., Dixon, T.H. and Engel, A.E.J. Reply.- GSA, Bull., 93, 174-178.

11. STACEY, J.S., DOE, B.R., ROBERTS, R.J., DELAVAUX, M.H. and GRAMLICH, J.W., 1980. A lead isotope study of mineralization in the Saudi Arabian Shield. Cont. Min. Petr., 74, 175-188.

12. FLECK, R.J., 1979. Rubidium-strontium geochronology and plate tectonic evolution of the southern part of the Arabian Shield. USGS Saudi Arabian Proj. Rept. 245.

13. HEIKAL, M.A.,1985. Major and trace elements in some granitic rocks southeastern At-Taif area, Saudi Arabia - their implications to magma genesis and tectonic setting. Szeged Acta Min.-Petr., 27, 5-16.

14. GINZBURG, A., FOLKMAN, Y., 1981. Geophysical investigation of crystalline basement between Dead Sea Rift and Mediterranean Sea. AAPG, Bull., 65, 490-500.

15. El-GABY, S., El-NADY, O. and KHUDEIR, A., 1984.

Tectonic evolution of the basement complex in the central Eastern Desert of Egypt. Geol. Rdsch., 73, 1019-1036.

16. STURCHIO, N.C., SULTAN, M. and BATIZA, R., 1983. Geology and origin of Meatiq Dome, Egypt: A Precambrian metamorphic core complex? Geology, 11, 72-76.

17. NAIRN, A.E.M., RESSETAR, R. and DAVIES, J.R., 1980. Paleomagnetic results from Pan-African rocks of the Egyptian Eastern Desert. Ann. Geol. Surv. Egypt, 10, 1013-1026.

18. BAKHIT, F.S., MELEIK, M.L. and El TAHER, M.A., 1984. Correlation of the radiometric survey data and the structural analysis with the geology of Abu Zawal area, central Eastern Desert, Egypt. Geol. Rdsch., 73, 833-851.

Part II

RESTLESS BORDER OF GONDWANA

The opposite sides of the Atlantic Ocean resemble each other like the edges of a torn newspaper. From this realization, came the idea of continental drift and its successor, plate tectonics. Originally, only one terrestrial landmass existed, the hypothetical Pangea. Perhaps due to convective movements in the mantle of the Earth, it started to split up about 300 million years ago. A northern supercontinent, Laurasia, separated from the southern one, Gondwana. They were divided by the Tethys Ocean which is now represented by the great mountain chains of the Alps, Atlas, Carpathians, Taurus, Alborz and Himalayas together with the Mediterranean, Black and Caspian Seas.

The regions forming the subject of this book lay in an unclear geographic location which could not have been entirely in Gondwana which incorporated South America, southern Africa, India, Australasia and Antarctica. However, the restless northern border of the southern supercontinent probably formed its southern margin.

Marine inundations and retreats occurred in relation to this. There were at least two glacial ages which left their imprints on the area and both were widespread in extent. The earlier Sahara Glaciation took place in the terminal Ordovician about 500 million years ago. The later Gondwana Glaciation was a Permo-Carboniferous event which happened in the southern hemisphere about 300 million years before the present. Their traces include tillites and allied phenomena.

Chapter II-3

EBB AND FLOW OF EPICONTINENTAL SEAS

"Quid magis est durum saxo, quid mollius unda? Dura tamen molli Saxa cavantur aqua".
Ovid, Ars Amat, I, 475.

CALIGINOUS PANGEA

"Pangea" is the name applied to a speculative supercontinent which included all continental crust prior to the Carboniferous. Even today, many earth scientists are reluctant to concede its former presence. However, two protocontinents, Laurasia to the north and Gondwana to the south, are accepted by most geologists as having existed from the Carboniferous to the Jurassic and may have derived by fragmentation of Pangea.

Unfortunately, confusion over terminology is evident in very recent publications. Thus, for instance, in one of these, Gondwana is assigned to the Visean and Pangea to the Early Permian It is proposed that, in Late Devonian times, Gondwana and its northern counterpart approached each other at a rate of 15 km per million years. During the Carboniferous, they are said to have merged to form a north-south oriented landmass. Later, in the Early Permian, Angara is alleged to have joined to create Pangea, a supercontinent envisaged as extending from the south pole to within a few degrees to the north pole and breaking up by the Jurassic. This is a clear misuse of the strictly defined term "Pangea" (1).

The basal shelf of southern Egypt comprises Precambrian to Early Paleozoic granites, gneisses and other igneous and metamorphic rocks over which clastics and carbonates were deposited during most of the Phanerozoic. Basement outcrops are common in many areas in southwestern Egypt from the Nile to Gebel Uweinat. North of Aswan, small outcrops are associated with fault lines of minor vertical displacements up to 100 m. Above this parallel, Pre-

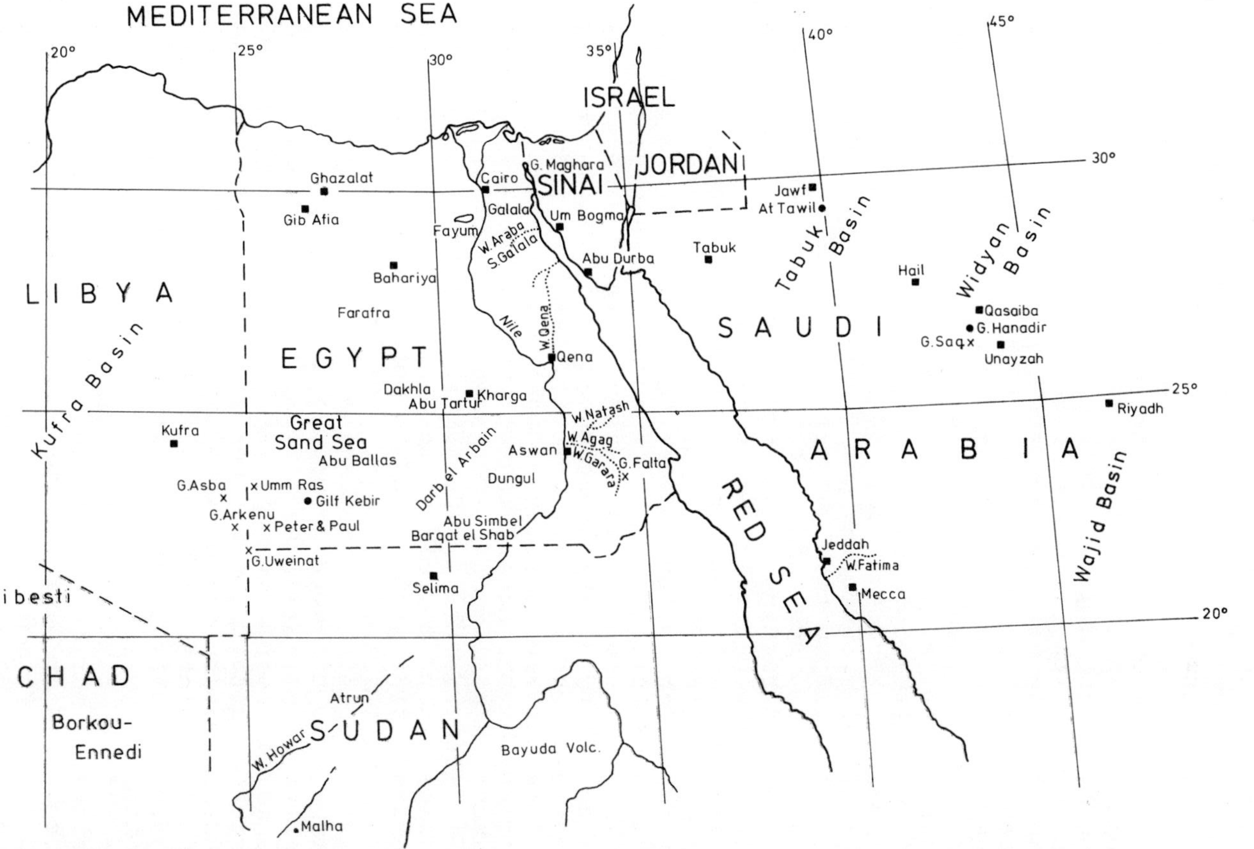

Figure 3.1: Locality map of region.

cambrian to Early Paleozoic igneous rocks are not exposed. The Gilf Sandstone at Gilf Kebir is unfossiliferous. It is identified as Jurassic and Cretaceous on the basis of its stratigraphy. Also, it is equivalent to fossiliferous beds of this age on and below the surface in the Kharga-Dakhla region (Figure 3.1). The Mesozoic sandstones in southwestern Egypt have been divided into seven units of which three were placed in the Jurassic and the rest in the Cretaceous. Consistent units have been claimed between Kharga and the Gilf Kebir, i.e. over a distance of 600 km, correlating primarily from Lingula and plant remains without mention of index fossils. However, this does not take into account facies changes and, in any case, proposed boundaries cannot be followed in the field. Structural contouring on the base of the Duwi or Phosphate Formation shows that two basins and three highs were responsible for individual centers of deposition ranging in age from the Paleozoic in the west to the Mesozoic in the east (2).

Existing continental remnants of Gondwana separated about 140 to 130 Ma in two different places, namely Africa from India-Antarctica and India from Australia-Antarctica, with accompanying development of the Indian Ocean. This ocean subsequently expanded by sea-floor spreading and drifting of the loosened continents. However, grabens and normal faults occurred earlier along the eastern margin of Africa which does not indicate oceanic activity, but rather a preliminary continental faulting. In the same manner, the marine Jurassic sediments of East Africa must be related to a southwardly directed Tethyan transgression on to Gondwana instead of the dawn of the Indian Ocean. The breakup of Gondwana and the opening of the Indian Ocean were associated with a continuous narrowing of the Tethys by the northwardly drifting India. The tectonic and magmatic movements of the last 140 Ma did not proceed continuously. Contemporaneity of tectonic and magmatic culminations imply a common origin of the movements deriving from geokinematics and perhaps indicating anomalies in deep mantle convection motions.

On the margin of the Arabian-Nubian Shield, Paleozoic marine and continental fossils were first recognized in Sinai. The construction of the Hejaz Railroad promoted work on the Paleozoic in Jordan and Saudi Arabia. Intensified interest in the Paleozoic of the latter arose from the granting of oil concessions to CASCO (Aramco) in 1933 and resulted in systematic stratigraphic investigations. A similar interest in petroleum was a major factor in further extending such activities into Libya and northern Chad. Toward the end of the last century, it was believed that all the thick and almost unfossiliferous sandstone sequences in southern Egypt were continental deposits of Cretaceous age. Even into the 1930s, it was impossible to differentiate Paleozoic from Mesozoic biostratigraphically in the sandy sequences. However, journeys to Gebel Uweinat by Ahmed Hassanein Pasha in 1923 and Prince Kemal el Dine Hussein in 1925 had already confirmed Carboniferous sediments by fossil plants from the extreme southwest of the country.

The term "Nubian Sandstone" was so controversial that a number of attempts have been made strictly to define it. At present, it appears appropriate to restrict it to Cenomanian and subsequent Cretaceous strata. Many data on both the Mesozoic and Paleozoic of southern Egypt were accumulated in the last decade. In the western Gilf Kebir, extensive Paleozoic sequences with possible Ordovician, definite marine Silurian and Early Carboniferous were indicated, whereas the Devonian was only suspected. However, the most interesting beds appear to be tillites and fluvioglacial sandstones which at first were convincingly correlated with the Late Ordovician Sahara Glaciation, but subsequently and controversially reassigned to the Late Carboniferous Gondwana Glaciation by the same author. Generally the Paleozoic clastics are separated from the igneous and metamorphic complex by a sharp disconformity. Nevertheless, due to both Paleozoic paleorelief and subsequent epeirogenic movements, stratigraphic sequences of different ages may be found in contact with basement in the various regions.

Cambrian is alleged to occur in Saudi Arabia, Jordan and Libya, but the evidence is ambiguous being based on ichnofossils such as Cruziana. Some workers consider this trilobite track to be Early Ordovician. In fact, the stratigraphic position of the relevant sandstone can correlate with any time from Cambrian to Early Ordovician. Disregarding Cruziana-bearing sandstones from Sinai as well as the Eastern and Western Deserts, definite Cambrian fossils like archeocyathids were found in surface exposures only in southern Sinai. Fossiliferous subsurface Cambrian is evident in some deep wells in the Western Desert. Two trilobite species were obtained from cores in a couple of these. The first is a partial cephalon of the Ordovician Protocalymene mcallisteri from Gib

Afia at 2,424 to 2,427 m depth and the other three specimens recorded from Ghazalat at 3,723 to 3,728 m depth and comprise incomplete cephalons of a Middle Cambrian Paradoxides.
The Early Cambrian of Sinai seems to correlate with the Fatima Formation on the Arabian Shield where stromatolites and archaeocyathids were also observed. This highly folded and faulted volcano-sedimentary formation rests unconformably on a 770 Ma old basement complex. The implication is that the final cratonization of the Arabian Shield occurred rather late in post-Fatima time, an observation supported by radioisotope data. K-Ar whole rock ages of 602 Ma and 586 Ma as well as 549 Ma have been measured, whereas Rb-Sr ages suggest a weighted mean of 688 Ma for the Fatima lavas. The K-Ar dates cover the Early Cambrian (Caerfai) time span, but the Rb-Sr ages appear to be rather old (Sinian) for the occurrence of archaeocyathids. It should be mentioned that the basement exposed in the Eastern Desert of Egypt furnishes similar ages. Principal extensional movements, such as the extrusion of the Dokhan volcanics, dyke emplacements and the intrusion of younger granites, took place during the Early Cambrian at 600 to 575 Ma BP. This corresponds with the Pan-African Orogeny (3,4,5).

THE ARABIAN PLATFORM

In Saudi Arabia, a broad belt of Paleozoic sediments curves around the Hail Arc and extends into the Tabuk and Widyan Basins. The sequence is mainly composed of sandstones and shales. It can attain a thickness of more than 1.5 km and ranges in age from Cambro-Ordovician to Permian. There are several pronounced stratigraphic breaks within the sections relating to the Caledonian and Hercynian Orogenies. The Cambro-Ordovician Saq Sandstone is up to 600 m thick and contains Cruziana at the top. It is overlain by Llanvirnian shales reaching 150 m thickness. Index fossils include graptolites, trilobites, chitinozoans and acritarchs. These shales are succeeded by a sandstone attaining a maximum thickness of 390 m with abundant Skolithos beds. Finally, Caradocian shales up to 95 m thick occur and yield graptolites, nautiloids and chitinozoans.
An eroded surface with up to 200 m thick conglomeratic channel-fill marks the contact with overlying sandstones. In the former beds, evidence of an

Ordovician glaciation is afforded by tillites, striated erratics and dropstones. A maximum age for these is obtainable from overlying thick Llanvirnian shales which are more than 200 m thick and contain graptolites and trilobites.
Shales with Climacograptus grade into sandstones which reach a thickness of 765 m and overlie the Silurian. Above them is a hiatus in the section evidenced by channel cuts on which were deposited continental sandstone up to 190 m in thickness. Early Devonian sediments were deposited again in an epicontinental sea with fluctuating salinity. This sequence commences with shales yielding palynomorphs. Two carbonate units with brachiopods and pelecypods have basal limestones and overlying dolomites together with a lagoonal, gypsiferous shale separating them.
Silicified trunks of Prototaxites are found to the west of Sakakah in northwestern Saudi Arabia, resting on transitional beds between the Early Devonian Jawf Formation and the overlying continental Sakaka Sandstone. The remains refer to huge arboraceous algae, which were adapted to an amphibious mode of life in an off-shore, brackish-marine environment. Such fossils were also identified from the Devonian of the Kufra Basin in Libya.
Continental Middle and Late Devonian rocks were also found in boreholes. There are no Late Paleozoic sediments preserved in the Tabuk Basin, but Late Carboniferous and Permian are known to the east of the Hail Arc within the Widyan Basin. Continental sandstones and shales of Early to Late Carboniferous age dated by sporomorphs and were only penetrated in wells, but Late Carboniferous to Middle Permian sediments progressively overlap older Paleozoic rocks. These are sandstones, shales and impure limestones, up to 180 m thick, which reflect continental to marginal marine deposition. Sporomorphs from well samples cover all ages from Westphalian to Kungurian. Near the base of the exposed section at Unayzah, abundant plant fossils with Annularia, Cordaites, and Pecopteris were detected and grew during Late Carboniferous to Early Permian time.
The Permian to Middle Triassic sequences of the Levant are characterized by a clastic-carbonate platform facies. As a rule, the succession of Permian to Middle Triassic rocks along the northern and eastern margins of the present Arabian Shield comprises a set of interdigitating sedimentary facies which record transgressions and regressions of the

Tethys.

Where present, the Early Permian comprises fine-grained clastics with secondary clastic-carbonates originally deposited in a shallow marine environment. A fossil flora in central Saudi Arabia confirmed the age of the deposits. The Late Permian is marked by a change in sedimentation from dominant clastics to mostly carbonates which begin with a widespread transgression over most of the Middle East extending near to the edge of the Precambrian-Cambrian basement of the Arabian Shield. The Tethys Ocean deposited variably textured carbonates of very shallow marine conditions over much of the Arabian foreland, including Oman. However, restricted basins developed periodically across the present Rub al Khali, the Empty Quarter, and a carbonate-evaporite sequence dominated. On the other hand, in the extreme south of Saudi Arabia and locally in uplifted areas, continental deposits were laid down. In central Syria, shallow basins formed and in these, highly fossiliferous thin-bedded limestones and argillaceous carbonates together with shales, neritic shelf deposits, accumulated (6).

Late Permian marls, dolomites and evaporites are restricted to the Widyan Basin and are not recorded from Jordan. In this latter country, peneplanation of deeply weathered plutonites took place at the beginning of the Cambrian. A distinct relief was carved in the Precambrian surface of Wadi Araba. In this period, the paleogeographic evolution of Jordan was dominated by the interrelationship between the Tethys Ocean and the Arabian-Nubian Shield. Continental sedimentation of the Early Cambrian was followed in the middle part of the period by a Tethyan transgression reaching the eastern rim of the central Wadi Araba. Regression followed with accompanying volcanism in this region of the Wadi. In the Ordovician, there was another transgression toward the south by the oscillating sea and sandstones and sandy shales deposited with occasional sandy shales and evaporites in lagoons and bays. By Late Silurian times, continental deposition of clastics recommenced. No further Paleozoic deposition has been recorded in Jordan.

In Sinai, Paleozoic sequences up to 230 m thick were usually assigned to the Carboniferous until it was shown that they contain the ichnofossils Cruziana, thus proving their greater antiquity. The thick sandstone section underlying the dolomites and ore deposits of the Early Carboniferous Um Bogma Formation resemble those exposed in the southern

Israeli and Jordanian Cambrian. Early Cambrian trilobites were encountered in Israel as well as the significant ichnofossils Cruziana and Rusophycus, but younger unfossiliferous sandstones can only be said to be pre-Carboniferous in age.

Figure 3.2: Paleozoic beds disconformably lying on Precambrian Basement at Um Bogma, Sinai.

The Um Bogma Formation consists of dolomites up to 43 m in thickness with basal ferro-manganese ore-bodies. These are important and productive and the most abundant ore mineral is ferruginous pyrolusite in various habits, but manganite is also common. The first stages of the manganese-iron replacement of the dolomite appear in narrow black streaks and spots in the latter. The radioactivity of replaced dolomite always increases progressively with the rise of pyrolusitization. A rather even distribution of uranium in the ore bodies implies that this metal was co-precipitated from solutions with these minerals. Main mineralization occurred in the Miocene, but there may have been renewed pulses of various hydrothermal activities later on. This is suggested by the alternation of faulting and mineralization, the rejuvenation of faults and the encrustations of turquoise or malachite on joint surfaces of clastics previously stained by iron or iron-manganese oxides (Figure 3.2).

About 8 km southeast of Um Bogma, at roughly the same stratigraphic level as the metasomatic manganese ores, there are turquoise deposits consisting of joint fillings or concretionary nodules. In the lower and middle part of the carbonate units which attenuate southeastward, marine fossils such as corals, bryozoans, brachiopods and molluscs are common. They are of Tournaisian to Visean age. This fauna and its paleotemperatures determined from Pachypteria indicate that they lived in a tropical, epicontinental sea. Marine Early Carboniferous has not yet been seen either in southern Israel or in Saudi In the Paleozoic section of Abu Durba which is 70 km south of Um Bogma, the Carboniferous is mostly represented by clastics. Sandstones from there yield impressions and casts of Carboniferous driftwood made up of bits and pieces of Lepidodendron, Sigillaria and Asterocalamites, the marls and limestones contain ledges of bryozoans, brachiopods, and pelecypods (7).

The Cambro-Ordovician 40 to 100 m thick Araba Formation is composed of medium and coarse grained sandstones which were deposited on an uneven basement complex. Skolithos beds and Cruziana tracks are common and indicate a marine, littoral environment. It is separated from the white, massive sandstones of the Naqus Formation by a distinctive paleosol. Quartz gravels, irregularly distributed in the sandy matrix, are particularly characteristic of the formation and could be dropstones of fluvio-glacial origin. If this is correct, these sandstones could be contemporaneous with the Late Ordovician diamictites in Saudi Arabia.

The Naqus Formation is succeeded by a 15 m thick bioturbated sandstone and clay unit with a paleosol and a basal conglomerate, the whole sequence is tentatively assigned to the Devonian and is topped by the Carboniferous. This latter starts with another paleosol and transgressional conglomerate. The Early Carboniferous is overlain by a thick, clastic sequence up to 200 m thick which comprises predominantly sandy rocks, these containing shales, clays and even thin coal seams in the lower part. Marine horizons could be traced by some foraminiferids suggesting a paralic, sedimentary environment. Carboniferous plant remains are quite common. Palynologic ages obtained from the coals are controversial ranging from Visean to the Moscovian Epochs. In the Gulf of Suez, Carboniferous Black Shales have been recorded overlying Paleozoic rocks of 400 m thickness, the so-called Nubia-C sandstone. This may be of Devonian age and constitutes an excellent oil reservoir rock. Deposition occurred in an embayment extending from Suez in the north to Hurghada in the south and was terminated by the Hercynian Orogeny. Subsequently the Mesozoic was unconformably deposited. The coarse-bedded sandstones of the Carboniferous Ataqa Formation are capped by an up to 100 m thick sequence of red and brown unfossiliferous sandstones together with varicolored clays and shales commencing with a paleosol. The typical lithofacies is reminiscent of Permo-Triassic beds from the northern part of the Eastern Desert.

THE NUBIAN PLATFORM

In Libya, thick and in part well-dated Paleozoic sections are exposed in the Ghadames and Murzuq Basins to the west and the Kufra Basin to the east of the Tibesti uplift. The Paleozoic of the Kufra Basin extends to Gebel Uweinat on the Egyptian border, comprises clastic sediments and has not been studied as well as in the northern basins. Lithostratigraphic units, often barren of fossils, onlap and are deposited in either continental or marginal marine environments. They represent a Paleozoic profile which developed similarly to that in the Murzuq Basin. Instead of shelly fossils, bioturbation was detected in many horizons of the arenaceous units, thus enabling a reliable indication of paleoenvironments to be obtained. It is also useful in separating the rock units on a parachronologic basis. The Precambrian Basement Complex is covered unconformably by thick sandstone sequences, often cross-bedded, partly conglomeratic and devoid of index fossils. However, in certain horizons, trace fossils such as Skolithos and Arthrophycus may occur frequently. At Gebel Arkenu, the clastics are subdivided into 200 m of Cambrian sandstones followed unconformably by 150 m of Ordovician conglomerates and sandstones which grade into pale or whitish, commonly cross-bedded sandstones towards the northwest. A Late Ordovician age is inferred from brachiopods such as Hirnantia or Plectothyrella.

Drill cores from the Murzuk and Ghadames basins of western Libya have yielded Late Ordovician sporomorphs from continental plants. Llandoverian to Wenlockian megafossils of lycophytes, and psilophytes also occur in Libya and support the conclusion that land plants, including vascular ones, probably had a long pre-Silurian record extending back at least to the basal Caradocian. The Cambro-Ordovician sandstones in Libya embody glacial features (8).

The Silurian transgression came in from the Homra Basin in the west and reached its maximum extent around Ennedi, the southern rim of the Kufra Basin in northern Chad, during the Llandoverian Epoch. There are two major rock units in Central Sahara, namely the Tanezzuft Shale, containing graptolites like Climacograptus and Diplograptus, and the Acacus Sandstone in which Monograptus has been found.

The Late Silurian has not been found in the southern basins of Libya. However, an interesting fossil flora from the Late Acacus Formation in the Murzuq Basin comprises psilophytes and lycophytes which suggest transitional Siluro-Devonian sequences. Devonian clastics have a wide distribution, mainly along the southeastern margin of the Kufra Basin. Resistant continental sandstones, often cross-bedded and generally with a basal conglomerate, initiate the Early Devonian cycle of sedimentation. Palynomorphs and marine shelly fossils confirmed this age. However, except in the Emsian Age, marine horizons are very rare in the eastern Murzuq Basin. Instead, fossil plants, especially the arborescent lycopods, indicate warm, humid continental conditions. In the southeastern part of the Kufra Basin, highly bioturbated sandstones of considerable thickness (160 m) represent the Middle to Late Devonian. The age is based on correlation with fossiliferous beds within the Murzuq Basin. Only a few casts of brachiopods and pelecypods were found in the northern Kufra Basin. These fossils record the outer limit of the Givetian transgression which reached as far as northern Chad. This epicontinental sea has left behind fossiliferous and bioturbated sandstones which indicate a once flourishing marginal marine benthonic biota. The area of deposition may well compare with extended tidal flats.

Western Libya was also inundated by a Carboniferous sea which extended far south to Chad and Niger. Marine sequences with fossils are well developed in the Ghadames Basin embracing Tournaisian to Moscovian Epochs. On its southern periphery, there is an extensive exposure of the Early Carboniferous Mrar Formation. The base of this comprises both a cap rock and possible hydrocarbon source for many

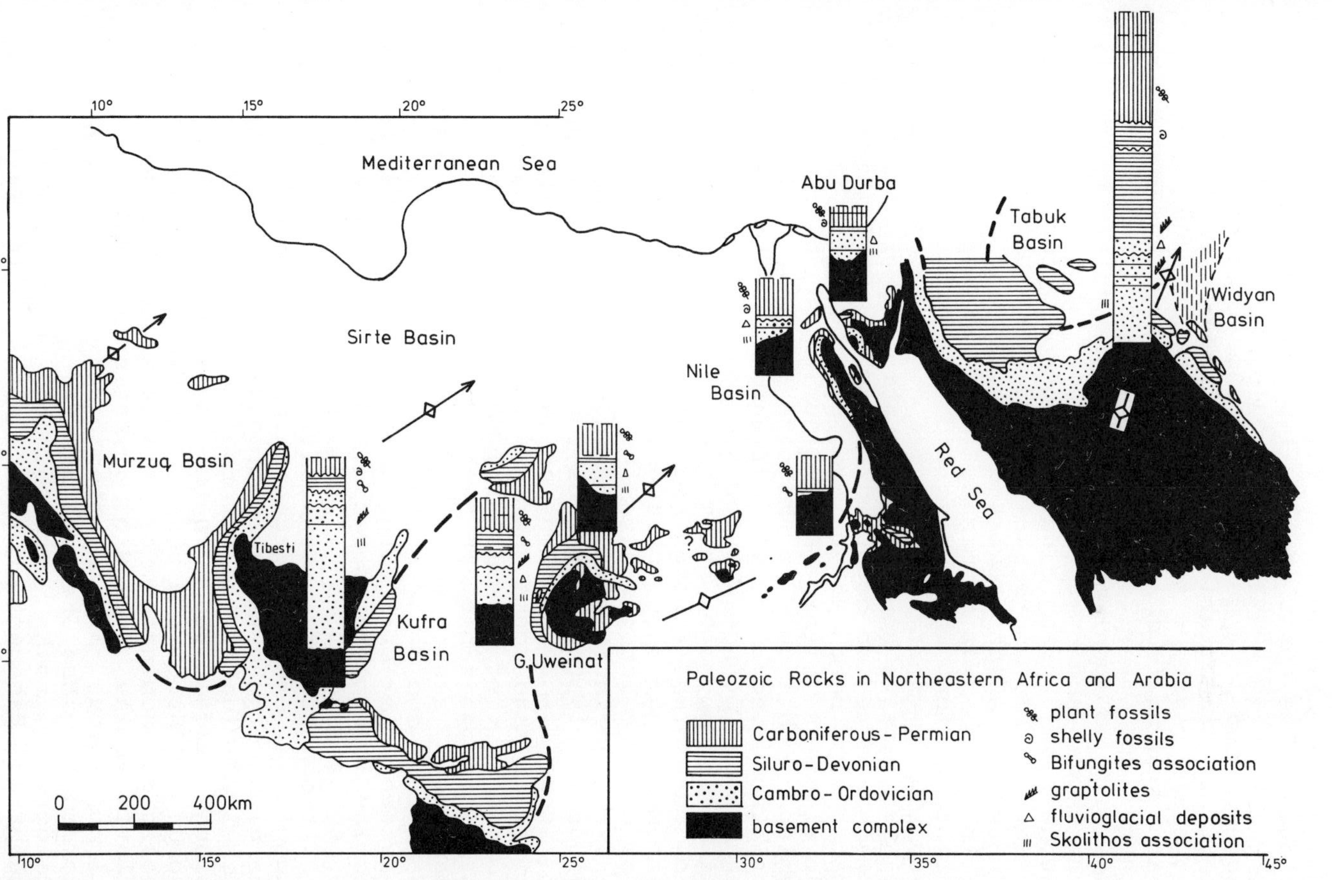

Figure 3.3: Paleozoic rocks in northeast Africa and Arabia.

producing reservoirs in the underlying Tahara Sandstones. The Mrar itself also contains significant gas shows southwest of the outcrop. The formation accumulated in a deltaic environment on the northern part of the stable Saharan Craton. It comprises several clastic sequences which coarsen upwards, each grading from shale to sandstone. In addition, there are many sedimentary facies ranging from deltaic to open marine shelf. The delta prograded westwards and was at first dominated by fluvial processes. During Middle Mrar deposition, it became progressively more influenced by shallow marine processes which reworked fluvial sand into bars. This phase was terminated by a widespread marine transgression which diverted the discharge regime some 100 km northward. In the higher Mrar sequences, a progressive southward migration of the diverted discharge regime occurred. An upward transition from bars into beach and barrier sands results from the development within a wave-dominated delta system which continued to migrate south, ultimately reintroducing fluvial sedimentation into the upper sandstones of the outcrop sections. The Mrar Formation is capped by stromatolitic limestones thought to be formed through extensive delta abandonment accompanying slow subsidence, high evaporation and low clastic imput (9).

In the northern Murzuq Basin marine sediments include Early Carboniferous invertebrate fossils such as corals, brachiopods and even goniatites. Continental sandstones increase towards the south, but, to the east of the Tibesti, marine Visean beds are still recognizable. In northern Chad, however, thick sandstone sequences (500 m) of possible Visean to Namurian age only contain plant remains. On the eastern rim of the Kufra Basin, no Carboniferous marine invertebrates were recognized, but there are bioturbated horizons, sometimes in sections with driftwood remains. Especially interesting are large specimens of Bifungites which might relate to the margins of the Visean epicontinental sea. Around Gebel Uweinat, the outcropping Carboniferous sandstones exceed 300 m in thickness and yield fossil wood and Stigmaria beds. A cratonic scenery of deposition dominated by fluviatile and sometimes fluvio-marine cycles of accretion is implied. Late Carboniferous and occasional Permian are only recorded from the Ghadames Basin.

Paleozoic rocks unconformably overlie the Precambrian Basement Complex in the Gilf Kebir and Gebel Uweinat region. Since abandonment of the comprehensive and easily understood term "Nubian Sandstone", unravelling the intricacies and interrelationships of its successors becomes ever more difficult. Far from further clarifying the stratigraphic situation, the waters of apperception have become muddied to the point of obscurity at which knowledge gives place to nomenclature. Thus, what follows is an attempt to shed light in dark places (10).

Most recently, the Gilf Kebir section was stated to contain 400 m of thick-bedded Silurian sandstones resting on the basement and containing Cruziana and Arthrophycus horizons. It composes the southeastern Abu Ras Plateau as well as its southern foreland. There are several Precambrian outcrops in the southern forelands of the Gilf escarpment, some of them showing onlapping, coarse, white sandstones above a basal conglomerate of 5 to 10 m thickness. In these sandstones erratic quartz pebbles occur.

Ordovician sandstones containing abundant Skolithos and also the indicative Cruziana rouaulti were found only further south in the Gebel Uweinat area also just above the basement paleorelief. Cross-bedded sandstones up to 70 m thick may correlate with an Early to Middle Devonian sequence in Libya. In Wadi Abd el Malik, there are 10 m of thick silty clays containing plant detritus. The overlying sandstones were highly bioturbated by the creatures producing Bifungites and Spirophyton.

The Carboniferous deposits are up to 100 m thick and consist mainly of marine shale and siltstone with fine-grained sandstone having fluvial beds in the upper part. In the center of Wadi Abd el Malik, the sequence contains a fossiliferous horizon with lycophytes which is followed by a marine succession indicated by ichnofossils, Camarotoechia and pelecypods. The identification of Bifungites fezzanensis is particularly interesting because this species cannot be distinguished from Libyan Late Devonian fossils. This is also true for species of Cruziana which resemble Devonian forms. Although large specimens of other Bifungites species are recorded from the Early Carboniferous in Libya, Cruziana has never been found in it. The question is which fossil genera provide a reliable chronology - widespread marine ichnofossils or continental plants? After all, Lepidodendropsis africanum and L. hirmeri together with species of Heleniella and Lepidosigillaria are described from the Early Devonian in the Murzuq Basin in Libya in a floral association which, despite its youthful Late Devonian to Early

Carboniferous appearance, is undoubtedly Early Devonian in age.

The eastern Gilf Kebir is also composed to a great extent of Paleozoic beds. However, Early Paleozoic deposits are less well exposed than in the Abu Ras Plateau because of a general northeastern dip of the beds. Doubtless the upper limits of the Carboniferous are still vague and the Paleozoic-Mesozoic interval not yet defined in the whole area. Here, there may be a long sedimentary hiatus resulting from the Hercynian Orogeny, but further south this may be filled. Nevertheless, western and eastern Gilf Kebir essentially comprise Paleozoic rocks mantled to the north and northeast by Mesozoic deposits. A reference level is marked by the "colorful" Abu Ballas Formation framing the plunging Gilf Kebir on its northeasterly trend toward Abu Tartur. The age of this fossiliferous and well-characterized formation is held now to be Early Cretaceous. It is not settled how or even if the Abu Ballas Formation replaces uppermost units in the stratigraphical profile of the northeastern Gilf Kebir.

During most of the Paleozoic, marine sediments were brought into the northern Sudan from the northwest through shallow transgressions of the sea. They alternated with fluvial episodes deriving mainly from the southeast and contributing interdigitating riverine deposits to the sequence. By the end of the Paleozoic, the Sudanese-Egyptian border comprised the axis of an extensive east-west uplift. From this, Paleozoic and Precambrian strata were eroded and transported southward into a continental basin. The process continued until the beginning of the Jurassic and was accompanied by east to west faulting and acidic to intermediate magmatism. In the course of the Jurassic, the region subsided and at its end, marginal strata were eroded. On both sides of the international border, northwardly transported fluvial and deltaic sediments grade into shallow marine beds of Cretaceous to Early Tertiary age. The typical sediment of the region is precisely that Nubian Sandstone, formerly so clear, but now under a cloud as a valid stratigraphic term. Its components were redeposited several times and in several environments. Among them are attenuated shales and paleosols sporadically distributed in the bulk of arenites. Practically none of these deposits is eolian, nearly all resulting from moist climatic conditions or marine transgressions (11).

On the west side of the Gulf of Suez in Wadi Araba,

exposures of the Late Paleozoic have long been known, fossiliferous horizons defining Carboniferous and Permian sediments within a clastic section. Proceeding southward into the Eastern Desert, Paleozoic sandstones outcrop in the Wadi Dakhl near the southern Galala scarps as well as further south. At the northern tip of the exposed basement, coarse sandstones and grits blanket the irregular Precambrian surface with succeeding green and purple siltstones and clay. Locally, these have abundant trace fossils in them. The most significant ones are Cruziana and Rusophycus attesting a Cambro-Ordovician age. These 10 m thick and easily recognizable beds are truncated by a paleosol and succeeded by sandstones and clays surmounted by white, brittle sandstones of 40 m thickness. The upper part of these latter contains impressions of Early Carboniferous driftwood below which are several horizons with abundant casts of pelecypods, rare gastropods such as Bellerophon and a few brachiopods including productids. Brown sandstone and varicolored shale totalling 5 m in thickness and partly bioturbated, with a Dictyoclostus coquina, complete the section above the driftwood horizon. The stratigraphic gap between the Cambro-Ordovician and the Early Carboniferous exhibited in this wadi is almost certainly related to Caledonian or at least early Hercynian movements and subsequent denudation.

To the southwest in Wadi Hawashia, the stratigraphic situation appears to be similar to that in Wadi Dakhl. Varicolored clay beds, sandstones and a basal conglomerate level the Precambrian relief and underlie white and brown sandstones of up to 80 m thickness. These sandstones are bioturbated and yield plant fragments, impressions of sphenopsid stems and fern fronds, as well as some fish bones. The unfossiliferous white sandstone may have a similar age and may be part of a prograding delta of an Early Carboniferous river system.

Outcrops of Paleozoic clastics occur in a long stretch of the Wadi Qena below marine Cenomanian beds. However, the stratigraphic sequence is rather discontinuous because of the structural relief and erosional pattern. Bioturbated Cambro-Ordovician sandstones underlie massive white, cross-bedded sandstones with occasional recumbent bedding, containing gritty and pebbly lag deposits. However, most conspicuous are sporadically distributed erratic quartz pebbles which may be dropstones like those found elsewhere in the Naqus Formation. In

Wadi Qena, this sequence is truncated. A prominent paleosol is preserved in the depressions of its uneven surface. Above, there may be Devonian, but there are undoubted continental Carboniferous beds evidenced by wood fragments of Asterocalamites and Lepidodendron. The Carboniferous sea may not have extended as far south as the central part of the Wadi Qena, but it surely controlled the base level both for fluviatile erosion and sedimentation much further south (12).

Figure 3.4: Angular unconformity between overlying Late Cretaceous iron-bearing sandstones with a basal paleosol and Paleozoic sandstones east of Aswan at Wadi Agag.

Precambrian basement ridges penetrate the Late Cretaceous clastic cover around Aswan and in the Etbai Desert. These sandstones attracted attention as early as the last century because of their marine horizons with the here exotic Inoceramus as well as angiosperm leaves. In fact, the Cenomanian transgression reached considerably further southeast than Aswan. Two coquina beds with Exogyra africana having attached serpulids and Polydora borings in a clastic sequence encroach on granite hillocks as far as 150 km southeast of Aswan. Late Cretaceous pelecypods can be collected from sandstones near Aswan on both banks of the Nile. Inoceramus balli occurs in and above the oolitic ironstone ores (Figure 3.4). These

are commonly associated with lag deposits, indicate intervals of interrupted clastic sedimentation and nearshore iron precipitation. The iron must have come from exposed, weathered substrate and lateritic soils. From fossil Unionacea such as Iridina, brackish, marginal marine environments and influxes of continental water were inferred.
Contrasting opinions on the stratigraphic position of the pelecypods in or above the exploitable Aswan ore sequence became more than academically interesting when bioturbated sandstone beds with abundant Bifungites were detected nearby. The ichnofossils came from an horizon only half a meter above the upper ore bed and were identified as B. fezzanensis. They are well known from the Devonian of Libya, having been recorded from the Middle Devonian of the Kufra Basin. The yougest occurrence of this widely distributed genus is Early Carboniferous. In the mining district, fossiliferous Late Cretaceous sandstones rest on partly reworked oolitic iron ores, the whole discordant to underlying bioturbated arenites of disputed age. Bifungites beds on top of autochthonous thin oolitic ore-bands have also been observed in gently folded sandstones exposed in a branch of a wadi 150 km to the southeast of Aswan. In this area, the clastic sequence consists of up to 250 m of mainly sandstones with some clay and tuff beds resting on the uneven and block-faulted Precambrian Basement Complex (13).
Basal gritty sandstones are succeeded by highly burrowed beds with Bifungites and other ichnofossils, themselves followed by an up to 100 m thick sequence of coarse, cross-bedded sandstones containing Carboniferous driftwood. Above this is a clay bed grading into a kaolinized tuff which may correlate with the Wadi Natash volcanics. In this case, the tuff, like the volcanics, would be Late Cretaceous in age. The sequence is concluded by 100 m of thick coarse, cross-bedded Taref Sandstone with conglomeratic lenses with large silicified tree trunks. Intercalated shales and siltstones yield impressions of angiosperm leaves.

REFERENCES

1. ROSS, C.A., ROSS, J.R.P., 1985. Carboniferous and Early Permian biogeography. Geology, 13, 27-30.

2. ISSAWI, B., 1982. Geology of the southwestern desert of Egypt. In: Desert landforms of southwest Egypt: A basis for comparison with Mars, ed. El-BAZ, F., MAXWELL, T.A., NASA CR-3611, 57-66.

3. BASAHEL, A.N., BAHAFZALLAH, A., OMARA, S. and JUX, U., 1984. Early Cambrian carbonate platform of the Arabian Shield. N. Jb. Geol. Pal., Mh., 1984, 113-128.

4. GETTINGS, M.E., STOESER, D.B., 1981. A tabulation of radiometric age determination for the Kingdom of Saudi Arabia. D.G.M.R. Misc. Doc. 20, 1-52.

5. DARBYSHIRE, D.P.F., JACKSON, N.J., RAMSAY, C.R. and ROOBOL, M.J., 1983. Rb-Sr isotope study of Latest Proterozoic volcano-sedimentary belts in the central Arabian Shield. J. Geol. Soc. London, 140, 203-213.

6. SHARIEF, F.A., 1982. Lithofacies distribution of the Permian-Triassic rocks in the Middle East. J. Petr. Geology, 4, 3, 299-310.

7. JUX, U., OMARA, S., 1983. Pachypteria sinaitica n. sp. - eine aufgewachsene austernähnliche Muschel aus dem Unterkarbon Aegyptens. Palaeont. Z., 57, 79-91.

8. GRAY, J., MASSA, D. and BOUCOT, A.J., 1982. Caradocian land plant microfossils from Libya. Geology, 10, 197-201.

9. WHITBREAD, T., KELLING, G., 1982. Mrar Formation of western Libya - Evolution of an Early Carboniferous delta system. AAPG, Bull., 66, 1091-1107.

10. ISSAWI, B., JUX, U., 1982. Contribution to the stratigraphy of the Paleozoic rocks in Egypt. Geol. Survey Egypt, paper 64, 1-28.

11. KLITZSCH, E., 1984. Northwestern Sudan and bordering areas: Geological development since Cambrian time. Berliner geowiss. Abh., A, 50, 23-45.

12. Van HOUTEN, F.B., BHATTACHARYYA, D.P. and MANSOUR, S.E.I., 1984. Cretaceous Nubia Forma-

tion and correlative deposits, eastern Egypt: Major regressive-transgressive complex. GSA, Bull., 95, 397-405.

13. ZAGHLOUL, Z.M., El SHAHAT, A. and IBRAHIM, A., 1983. On the discovery of Paleozoic trace fossils Bifungites, in the Nubia Sandstone facies of Aswan. Egypt J. Geol., 27, 1-2, 65-72.

Chapter II-4

EPISODES OF GLACIATION

"Throned in his palace of cerulian ice, here Winter holds his unrejoicing court".
James Thomson, The Season - Winter, 1726

WHY ICE AGES?

In the Late Pleistocene, the Earth endured the latest ice age impact, landscapes being remolded by slowly moving glaciers and temperatures dropping with the surfaces of many parts of the world being depressed by the unimaginable weight of thrusting ice sheets. Simultaneously, vast quantities of water were withdrawn from the oceans to sustain the colossal ice masses with the result that the sea-level everywhere fell by well over 100 m and exposed large areas of the continental shelves as land. Although the scars of these cataclysmic happenings are all over the globe for everyone to see, erratics were for long believed to have been transported by the tremendous water currents of the biblical deluge. It is through Louis Agassiz and his work on the glaciers of Chamonix that the glacial theory triumphed over the preceding diluvial one. A friend wrote to him in 1841 saying that "you have made all the geologists glacier-mad here, and they are turning great Britain into an ice house".
It is now common knowledge that many ice ages have occurred sporadically through geologic time. The problems are how they come about, what stimulates the growth and expansion of ice sheets, why they retreat and, above all, will they return? Needless to say, many attempts have been made to answer these questions. Some failed because too narrow a view was taken without recognizing that all the mobile elements of the Earth, the oceans and the atmosphere as well as the ice, interact like an engine and therefore have to be considered if the air-sea-ice

system is to be explained. The engine runs on solar energy received everywhere. Some is reflected back into space and some is absorbed. The former relates to the terrestrial albedo which is 0.64 cal/cm², although this varies from land to sea and, of course, with the presence or absence of ice. At the 40th parallels north and south, there is an equilibrium between gains and losses. In all other places, the planet tends to heat or cool and the radiation budget is not in balance. Near the poles, snow and ice increase the radiation of energy back into space and there is a net loss of heat coupled with a very low sun angle. If only reflection and radiation operated, the poles would grow steadily colder with a correspondingly steady heating up of the equator. This does not happen because of winds and oceanic currents which transport equatorial heat poleward. To account for the ice ages, whether Pleistogene or earlier, the effects of an enlarging or diminishing ice sheet must always be recognized and so too must the correlation between its volume and the prevailing sea-level. When such a sheet is expanding, heat is lost by reflection and the global temperature falls so that more ice results. On the other hand, when contraction takes place, the temperature rises and the process is thereby accelerated. This may be called a radiation-feedback effect.

An early suggestion was that an ice age may be triggered by a decrease in the solar output of energy. There is certainly a connection between sunspot activity, rainfall and temperature, but, unfortunately, it has never been shown that a similar parallel exists with variations in solar energy. Thus, the influence of the sun on climate does not prove that solar fluctuations promote ice ages.

A second hypothesis is that the uneven distribution of dust in space may activate an ice age when the planet traverses a region where dense concentrations of it reduce the sunlight to such a degree that a cooling trend is initiated. An opposite version of this idea is that a warming trend may result from dust falling into the sun and causing it to emit more light and raise terrestrial temperatures. Obviously, these two propositions would have to be reconciled before adoption of either is feasible. The difficulty is that, as with the previous suggestion, there is absolutely no evidence whatsoever for either of them.

Another cosmic approach involves the astronomical

theory which takes into account the precessional cycle and variations in the shape of the orbit of the Earth. It predicts that one hemisphere or the other will undergo an ice age whenever there is a winter solstice far from the sun together with a markedly elongated orbit. Meteorologists objected that variations in solar heating entailed by the theory would be too small to affect climate. An extension and improvement of the idea is provided by the radiation curves of Milankovitch which, although the ice age timing and radiation minima do not fully agree, apply very well to the date of the last of these latter 25,000 years ago. Unexpectedly, the radiation curves correlate rather well with the isotopic curves obtained from deep-sea cores. For the Milankovitch effect, there is an oscillation period of about 41,000 years determined by variations in the tilt of the spin axis of the Earth. In a similar, so-called Croll effect, two oscillation periods exist, one of 23,000 years which is the precession of the equinoxes and the other of 100,000 years. This latter is the result of changes in the gravitational effects of the planets on the terrestrial orbit. Whereas the Milankovitch effect on the tilt in both hemispheres is the same, the Croll effect behaves in opposite ways in the northern and southern hemispheres. In addition, the periodicity of the former is about twice that of the penultimate value for the latter (1).

A more plausible concept is that variations in the quantity of carbon dioxide in the atmosphere may induce changes in the heat balance of the Earth. At present, the gas constitutes 0.03% of it. While reasonably transparent to the solar shortwave radiation, it is rather opaque to reflected longwave radiation, returning it to space. The more carbon dioxide there is in the air, the more the atmosphere acts like the glass which covers a greenhouse, this so-called "greenhouse effect" heating the environment through the trapping of incoming radiant energy. Some believe that if the concentration of this gas falls far enough, an ice age will result. Whether such an effect actually can occur is highly questionable as is its opposite, the so-called "icehouse state". This latter may be that in which we live.

A syntagma for two Phanerozoic "supercycles" has been presented. Each is supposedly 300 Ma long and both are claimed to reflect dynamic convection cycles in the mantle. Both are presumed to have had a first phase of rapid convection with many plumes that broke up Pangea-like supercontinents, raised sea-levels and flooded blocks of land. Intense volcanism released carbon dioxide from the mantle, while atmospheric losses of this gas to the lithosphere through weathering dropped together with the area of land. All this caused the greenhouse climates which characterized two major time intervals, namely from the Late Cambrian to the Late Devonian and from the Early Jurassic to the Late Eocene. A second phase of the convection cycle was slower and the number of active plates diminished with continents accreting and sea-level dropping. Volcanism declined, thus contributing less carbon dioxide to the atmosphere and the growing area of land took more out of it through weathering. Consequently, this gas attained a new equilibrium at a lower level in the atmospheric-hydrospheric sink. The result was that the greenhouse climate was replaced by an icehouse state marked by cold, dry polar regions, sea ice and continental ice sheets plus oceans which were both cold and highly oxygenated. These conditions seem to have existed around the end of the Proterozoic, in the Permo-Carboniferous and in the latter half of the Cenozoic, including to-day.

It could be that volcanism is a possible generator of ice ages for another reason. This is, that during frequent eruptions, fine volcanic dust concentrates in the atmosphere and reflects more solar energy back into outer space with result that the climate may cool. This is susceptible to examination. In 1883, Krakatoa injected such a mass of dust into the heavens that sunsets everywhere became redder for the next two years. Measurements showed that the average global temperature really dropped during this period. When the dust finally dissipated and settled to the ground, the climate became normal again. While, theoretically, it ought to be possible to test the volcanic dust theory by comparing historical records of ice age climates with sedimentary records of volcanism, there are not enough measurements over a large enough area to do so. Clearly, both the carbon dioxide and dust emissions could exert a combined climatic bias toward the greenhouse condition.

Still another model envisages that an ice age may begin when, for a short time, the Arctic Ocean is ice-free and open to warm currents from the North Atlantic. During such a time, evaporation would increase and the overlying air would acquire excess water vapor producing more snowfall on the surrounding lands. Escalating glaciation could then lead to an ice age, deglaciation commencing when temperatures would have dropped sufficiently to permit the Arctic Ocean to refreeze. Thereafter, with the moisture source removed, the ice sheets would shrink, the sea-level rise and warm North Atlantic currents could again start melting Arctic sea ice. The conception can be tested. This is because it outlines an historical sequence of events which should be recorded in Arctic Ocean sediments. Layers of these deposited at the inception of an ice age ought to contain fossils of organisms which colonized sunlit waters. However, no such fossils have been found and furthermore the sediments demonstrate that the Arctic Ocean has not been free of ice during the past several millions of years.

Figure 4.1: Sahara (coarsely dotted) and Gondwana (stippled) Glaciations with paleopoles. The Ordovician one is in North Africa, that of the Permo-Carboniferous in Antarctica.

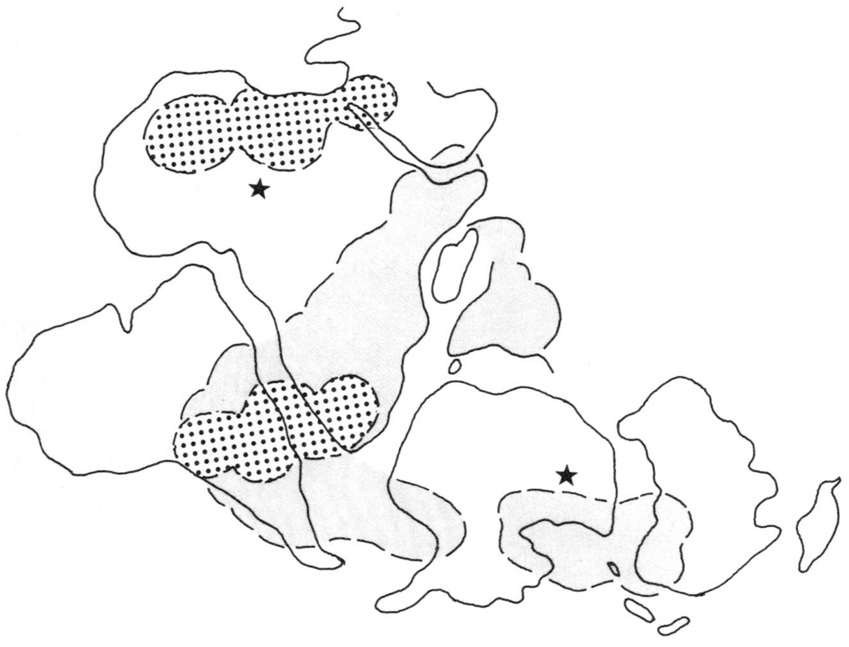

A development of this notion is designated the coincidence hypothesis and entails an open-sea north pole region surrounded by land and counterbalanced by a landmass covering the south pole. Plate movements could be the operative mechanism both for the Pleistogene and Permo-Carboniferous glaciations. If the idea is valid, it follows that a high concentration of land in high latitudes must perforce occur during ice ages. When such a continent with high ground temporarily incorporates the pole, ice begins to lie permanently on elevated regions and spreads gradually to create an ice sheet (Figure 4.1).

Finally, allusion should be made to the amusing stochastic or "no cause" theory according to which, no special explanations are necessary. Ice ages are regarded as just instances of large scale variations in the terrestrial climate which result from the accumulation of numerous small and random alterations in weather stored in oceans and ice sheets. There is no doubt that such random variations take place on monthly or yearly bases. Proponents of the concept suggest that the longer the time interval examined, the greater will be the consquent climatic variation. The flaw is that no empirical investigation is possible (2).

PRE-QUATERNARY GLACIATIONS

When the Proterozoic Eon began 2.5 billion years ago, the Earth was in a cooling cycle which peaked about 500 million years later in the first of an irregular series of glacial episodes of which the last, still continuing today, is discussed in chapter V-12 below when that of the Pleistogene will be examined. This Early Proterozoic glaciation left its traces in the form of tillites in Canada, Southern Africa and India, demonstrating a vast inland glaciation not long after the transition from the presumably warmer Archean. During the succeeding 1.5 or more billion years, continental glaciers were either rare or absent. However, in the Late Proterozoic, 850 to 800 Ma ago, another glaciation took place and its effects were so widespread that its glacigene sediments and striated pavements can be found on all the existing continents except Antarctica. Thus, this constituted the greatest of all ice ages and may even have ranged in age from 1 billion to about 600 million years, although several discrete glacial phases may be involved. At least three are known to have occurred in Africa, the last some 600 Ma ago, when glacial phenomena are recorded all over the world. Restricted paleomagnetic data presently available imply that there was some continental glaciation quite close to the current equator. Consequently, it is very likely that most of the planet was rather cold in Late Proterozoic times. Since glacial indications are very common in West Africa, there can be no doubt that the northeastern region and the Middle East were also affected.

During the Late Ordovician and Early Silurian, the Central Sahara experienced the maximum consequences of the third period of glacialism in earth history when the region lay near the south pole and a mass extinction occurred in the marine realm. This extermination of biotas may have been connected with a substantial drop in sea-level. Ancient moraines have been found together with striated boulders and erratics. All this spectacularly confirms the paleomagnetic data implying a North African pole position during this period. The relevant traces of glaciation have only conclusively been observed at high paleolatitudes in northern and southern Africa so that the glaciation seems to have been restricted to the polar region. There is no evidence of multiple glaciation as in other ice ages. It is possible that this so-called Sahara Glaciation might have resulted from some extraterrestrial catastrophic event which led to the great faunal crisis. Alternatively, this cryospheric phenomenon may have been triggered by an overshoot of the "greenhouse effect" promoting a dense cloud cover which raised the planetary albedo to increase reflection of solar energy and thus cool the Earth. It is of particular interest that the Cambro-Ordovician sandstone units in Libya commonly expose such typical erosional features of glaciation as sandstone pinnacles and blades, rounded peaks, U-shaped and overhanging valleys especially in the southeastern part of the Kufra Basin. This exhumed "Antarctic scenery" was covered by conglomerates, sandstones and shales attributed to the Late Ordovician Memouniat and Early Silurian Tanezzuft Formation (3).

Near the Hail Arc in Saudi Arabia, poorly-sorted conglomerates in the Ordovician contain huge boulders with striated erratics. These probably derived from local glaciations and ice rafting. A maximum age for these presumed tillites and dropstones is given by Llanvirnian graptolites and trilobites in the Qasaiba Shales which range from 35 to 210 m in thickness (4).

Possible Silurian tillites have been identified in North and South America which may point to a continuation of glaciation beyond the Ordovician. By the Early Silurian, plate movements shifted the glacial centers into southern Africa and South America. From the Middle Silurian, the planetary climate ameliorated until the Late Devonian, when glaciation possibly affected parts of Africa again. Thereafter, the glacial centers shrunk until the advent of the Permo-Carboniferous ice age, which culminated in southern Africa. The migration of glacial centers closely approximates paleomagnetic polar wandering paths (5).

In frigid Gondwana, during the Late Carboniferous, the extreme temperature differences between equator and poles were reflected by the northward thrust of continental glaciers. Coal deposits arose from a Glossopteris flora adapted to cool climates contemporaneous with tropical coal swamps in North America and Europe. Tethyan Ocean currents, supplemented by those of the Pacific, probably supplied enough moisture to different parts of Gondwana to have nourished the ice sheets. The Permo-Carboniferous Glaciation is manifested in the southern protocontinent, but probably not in North Africa.

On the southeastern margin of the Arabian Shield, there are continental sandstones known to precede the Late Permian and containing boulder beds thought to be glacial in origin. Thus, their stratigraphic position may be Permian or possibly older. Probably they were transported by rivers. If the beds contain real erratics, they must relate either to the Permo-Carboniferous Glaciation or to the earlier Ordovician one. Glacial diamictites interbedded with rhythmites are found also in Yemen and are termed the Akbra Shales; they comprise tillites and fluvioglacial deposits. Associated gigantic erratics and Precambrian basement striae record a glaciation dated as Permo-Carboniferous by pollen analysis from the rhythmitic shale sequences which gave a confirmatory age. A possible equivalence between the Akbra Shales and the Upper Wajid Sandstone of neighbouring Saudi Arabia may be inferred from the palynologic data. Northwest of Djibouti, near to Asmara, there are glaciogenic sediments, as in Yemen. These represent the northernmost such deposits in Africa except for those alleged to occur in the Gilf Kebir (6,7).

The principal phase of the Gondwana Glaciation extended from the Late Carboniferous to the Early Permian and continental ice sheets existed from the Andes to Australia, though whether or not contemporaneously is not known. Paleomagnetic evidence indicates that the pole was located in Antarctica during Stephanian-Sakmarian time. Some of the ice centers were situated upon uplands reaching altitudes as high as 1,500 m. All the glacial and periglacial facies identified during the Pleistocene glaciation, except loess and interglacial soils, have been recognized in Gondwana. The final, Early Permian deglaciation probably proceeded from South America over Africa to Antarctica, where glaciers persisted into the Middle Permian (8).

Figure 4.2: Glacial erosion at Peter and Paul Mts, Gebel Uweinat. An U-shaped valley in metamorphic basement can be seen. It is partly veneered by Early Carboniferous sandstones with fossils.

In the extreme southwest of Egypt at Peter and Paul Mountains to the northeast of Gebel Uweinat, the Precambrian basement reveals a remarkable geomorphology comprising rounded peaks, ridges and U-shaped valleys. Sandstones and gritty layers are preserved in depressions of the basement. Abundant Carboniferous plant fossils such as stems of sphenopsids, lycopsids, fronds of ferns and rhizocretions were recognized in these sandstones. Such fossiliferous sandstones cover parts of a U-shaped valley at Peter and Paul Mountains. Consequently, the present scenery is a product of actual desert weathering on a partly exhumed land surface which, of course, also has been subjected to volcanism. Similar morphologic features were observed in the adjacent Kufra Basin around Gebel Asba. Certainly, these Libyan and Egyptian paleoreliefs document the same mode of abrasion and probably a simultaneous cold phase of the Late Ordovician glaciation. From the inland ice, which must have covered vast areas, fluvio-glacial deposits were liable to accumulate and these may be represented by the Naqus Formation. Continental Paleozoic to Mesozoic clastics of southern Egypt and northern Sudan were deposited in a river system which ran south. The oldest beds derive at least partially from a glacial or periglacial environment. Varvites of a large glacial lake have been preserved in northwestern Sudan at the rim of the Kufra Basin and east of Gebel Kissu, but these are said to relate to the Permo-Carboniferous glaciation. The Late Jurassic to Early Tertiary sedimentary cycle of Nubia is a result of structural movements which were caused by the breaking up of Gondwana.

In the Wadi Abd el Malik on the Abu Ras Plateau at Gilf Kebir, there is a 50 m thick diamictite with an overburden of marine shales and sandstones extending up several hundred meters and said to yield Silurian graptolites at the base. There are shelly fossils and plant fragments in the middle and upper parts on the basis of which a Devonian to Carboniferous age may be inferred. Consquently, the diamictite could reflect the Late Ordovician Glaciation if it turns out to be a tillite. It was assumed that the diamictite interdigitates the Ordovician-Silurian Naqus Formation. This is a sequence of more than 100 m of white sandstones capped by shales with ichnofossils such as Cruziana. The formation has been suggested as being of fluvio-glacial origin, not least because of its erratic pebbles and also because of synchronous glacial phenomena in Libya and Saudi Arabia (9).

The supposed Carboniferous glacial sediments of Wadi Abd el Malik would have to be equivalent to the Dwyka Glaciations of South Africa if they are eventually found to belong to this system and also if they can be established as tillites rather than just mixtites. Of course, if the diamictite could be established as a debris avalanche rather than merely a tillite, the matter would be easily resolved (10).

During the Early Carboniferous, the pole was in the Transvaal.The famous Karroo sequence in South Africa includes the Late Carboniferous Dwyka Tillite deposited partly in a marine environment. In its upper part, there is a non-marine carbonaceous shale in which Mesosaurus is found. Also, other shales of this age contain the Glossopteris flora extending into the Permian Ecca Series of clastics containing coal seams. There are many paleogeographic and biogeographic arguments against any Dwyka equivalent ever being found in Egypt. It may be germane to add that no elements of the Gondwana biotas have ever been found in northeastern Africa despite exhaustive geologic work. The Gondwana Glaciation reached its maximum extent in the Late Carboniferous, the pole then being in western Antarctica. At the same time, part of the main ice mass covering southern Africa stretched into the Parana Basin in South America and also mantled most of Antarctica itself together with Tasmania and South Australia.

REFERENCES

1. HOYLE, F., 1982. Ice, 209 pp., NEL Books, Sevenoaks, UK.

2. IMBRIE, J., PALMER IMBRIE, K., 1979. Ice Ages - solving the mystery. 224 pp., Macmillan, London, UK.

3. BELLINI, E., MASSA, D., 1980. A stratigraphic contribution to the Paleozoic of the southern basins of Libya. In: The Geology of Libya, ed. SALEM, M.J. and BUSREWIL, M.T., 1, 3-56, Acad. Press, London, UK.

4. McCLURE, H.A., 1978. Early Paleozoic gaciation in Arabia. Palaeogeography, Palaeoclimatology, Palaeontology, 25, 4, 315-326.

5. CAPUTO, M.V., CROWELL, J.C., 1985. Migration of glacial centers across Gondwana during Paleozoic Era. GSA, Bull. 96, 1020-1036.

6. KRUCK, W., THIELE, J., 1983. Late Palaeozoic glacial deposits in the Yemen Arab Republic. Geol. Jb., B 46, 3-29.

7. KLITZSCH, E., 1983. Paleozoic formations and a Carboniferous glaciation from the Gilf Kebir-Abu Ras Area in southwestern Egypt. J. Afr. Earth Sci., 1, 17-19.

Glacial episodes

8. MARTIN, H.,1981. The Late Paleozoic Gondwana Glaciation. Geol. Rdsch., 70, 480-496.

9. KLITZSCH, E., 1979. Zur Geologie des Gilf Kebir Gebietes in der Ostsahara. Clausthaler Geol. Abh., 30, 113-132.

10. KLITZSCH, E., LEJAL-NICOL, A., 1984. Flora and fauna from strata in southern Egypt and northern Sudan. Berliner geowiss. Abh., 50, 49-79.

Part III

SHIFTING TETHYAN SHORES

The extensive seaway of the Tethys Ocean existed since the Late Paleozoic 250 million years ago, later separating the enormous northern and southern supercontinents and surviving in fragmentary form today, mainly as the Mediterranean Sea and, of course, the mountain ranges of the Old World. At the end of the Mesozoic, approximately 65 million years ago, India moved rapidly north across it toward Asia and, at the same time, Australia with Antarctica split off from Africa.

In such mountain belts, igneous ultrabasic rocks called ophiolites are found and resemble oceanic crust or uppermost mantle. There is an important set running from Oman, through the Zagros and Taurus, probably representing a former Tethyan subduction zone between Africa and Eurasia.

Mesozoic sediments all over the region provide traces of marine movements, shifting shores of the Tethys. Rich fossil faunas are found, often including fish such as sharks and reptiles like turtles and crocodiles. Important economic deposits of phosphate and coal, the "portable climate" of Ralph Waldo Emerson, accumulated.

The Cretaceous-Tertiary boundary cuts off dinosaurs, ammonites, belemnites, coral-like rudistids and inoceramids as well as many other forms from later biological assemblages. It was a time of mass extinction and excess iridium built up in clays, interpreted by some as the record of a bolide impact. Whether this is true or not, the Syrian Arc was initiated by the Laramide mountain- building episode and involved Cretaceous to Tertiary strata from the Egyptian desert oases through Sinai into the Levant as far as Palmyra. The story of this crescentic mobile zone is described in this section.

Chapter III-5

CLASTICS AND BIOGENICS

"Everything has its beauty, but not everyone sees it".
Kung Fu-tze, 557-479 B.C.

INTERMENT OF THE CRATON

In Saudi Arabia, non-marine to littoral clastics with thin marine limestone and evaporite beds, altogether 750 m thick, represent the Triassic Period. Early Triassic deposition indicates a wide regression. The Middle Triassic consists of non-marine and intermittent near-shore sandstones with some shales in the extreme south of Arabia. There, cyclic sedimentation indicates phases of Tethyan advance and retreat. Shallow marine carbonates of Triassic age are also exposed in Oman, southwestern Iran, Iraq, western Jordan and Sinai. The Levantine Triassic comprises four sedimentary cyles. Regression is indicated by carbonate-evaporite rocks at the top of the sequence in these areas and may reflect tectonic movements. From the Middle to the Late Triassic, there is a sharp differentiation of depositional facies patterns. In Syria, also probably in Lebanon and northern Oman, carbonate shelf deposits of very shallow marine conditions are found. A carbonate-evaporite sequence of intertidal and supra-tidal deposits accumulated in Jordan and southwest Iran. Close to the edge of the Arabian Shield, a belt of continental sediments, the over 315 m thick Minjur Sandstone, was deposited. Uplift and non-deposition were recognized in Sinai, the western coast of the Persian Gulf and southern Oman. Most interesting are rapid thickenings and abrupt lithologic changes of huge amounts of richly fossiliferous clastic, neritic deposits which accumulated in central and eastern Iran. A flysch-like sequence went on depositing in the extreme northeast and east of that country (1,2).

Fossiliferous Jurassic around Gebel Maghara in northern Sinai comprises an alternation of marine and continental deposits. The latter are cyclothems and contain plant remains. One of them, the Bathonian Safa Formation, is coal-bearing. A dozen thin coal seams were found by drilling and in poor outcrops. The coal is sub-bituminous with good coking properties. The main seam is about a meter thick and belongs to the Flame- or Gasflame-coal of the German classification. It contains a fairly large amount of resinite and resembles sapropel coals, an interpretation which may be inferred from algal colonies similar to the recent Botryococcus braunii. The reserves in the Safa Formation exceed 30 million tons. The whole sub-Cretaceous section is more than 1,500 m thick and over half is made up of carbonates, with one third clays and the rest sands.

Figure 5.1: The Jurassic Mashaba Formation in the Gebel Maghara anticline, Sinai.

The marine fossils belong to the European faunal province and have little affinity with that of East Africa. The ammonites found include the Late Bajocian assemblage containing index species such as Lissoceras coliticum and Spiroceras bifurcata. A comparison may be made with the much richer fossil fauna of the Middle Jurassic Dhruma Formation and the Late Jurassic Tuwaiq Mountain Limestone near Riyadh, Saudi Arabia. This includes corals, brachiopods and molluscs including ammonites. This limestone is an oil-producing formation in eastern Saudi

Arabia. The lower part is composed of biomicrites as is the upper, but the latter contains abundant corals, stromatoporoids and algae. The microfacies reflects successive episodes of deposition, the first in a calm sheltered sea floor which was subject to periodic agitation and the second in open sea conditions (Figure 5.1).

The Musandam Peninsula is the northernmost extremity of the Oman Mountains. It comprises an allochthonous sequence of Permian to Middle Cretaceous shelf carbonates which are separated from ophiolites to the south by a NE-SW trending belt called the Dibba zone. This is structurally complex and composed of allochthonous slope and basin facies sediments as well as Haybi volcanics, "Oman Exotic" limestones, subophiolitic metamorphics and ultramafic slices.

Correlation of lithofacies confirms that the breakup and rifting of an enormous east-facing carbonate platform occurred in Middle to Late Triassic time and this resulted in the formation of a shelf edge and a small ocean basin. The periphery was characterized from northwest to southeast by ooid-skeletal small bioherms on the shelf edge, a bypass foreslope of laminated ooze and basin margin accumulation of carbonate turbidites or debris flows. From Middle Triassic to Late Jurassic time, the shelf edge, which now coincides with the northern boundary of the Dibba zone, appears to have been stable.

In Israel, ammonites of Late Jurassic to Early Hauterivian age have been found in gray shales of the Negba-Heletz region of the southern coastal plain. The subsurface geology of Ayun Musa in western Sinai is Mesozoic overlain by fossiliferous Early Miocene sandstones and marls. Beneath the Neogene is a major unconformity below which is an interdigitating sequence of "Nubian Sandstone" and Middle and Late Jurassic marine clastics and limestones. The Bathonian incursion was preceded by paralic swamp deposition including coal seams up to 1 m thick recorded from wells at depths of 550 m. The sea retreated during the Rhaetian-Liassic in Jordan and was succeeded by another Tethyan transgression during the Dogger. In the Gulf of Suez, Jurassic fluvio-marine and marine sediments indicate that a shallow branch of the Tethys penetrated as far south as Wadi Araba.

Many wells have been drilled in the Western Desert penetrating Early Jurassic sediments which are usually sandstones, with occasional interbedded limestones and dolomites - in fact, limestones often cap them. The Middle Jurassic is made up of an alternation of shale, sandstone and limestone which are so poorly fossiliferous that no biostratigraphic zonation has been made of them. The Late Jurassic is normally represented by marine and reefal facies comprising dolomitic limestone which grades down to marly and shaly beds, the latter believed to be Callovian in age. The dolomites are overlain abruptly by the Early Cretaceous littoral to epi-

Figure 5.2: Stratigraphic cross-section of the Western Desert, Egypt.

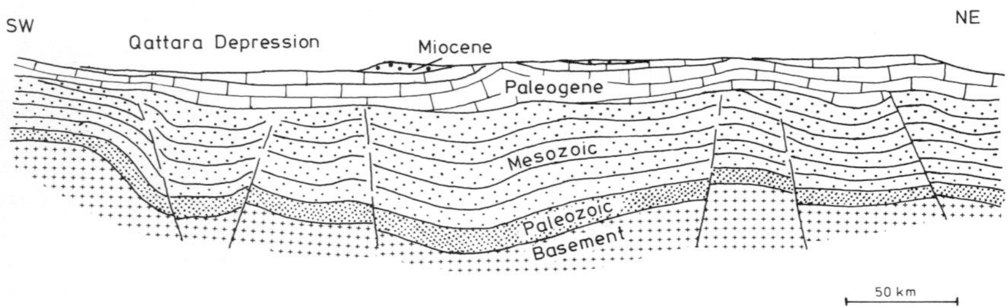

continental sandstone of Nubian facies. The Late Jurassic possesses index fossils which have facilitated biostratigraphic zonation. In general, Jurassic sediments become epicontinental towards the south, the furthest extent of the invading Jurassic sea. Further north, shallow marine clastics and reefal facies disappear and are replaced by deep water marine facies with shales, e.g. black shales in the Nile Delta (Figure 5.2).

On the eastern side of the Murzuk Basin, the sediments contain plant fossils such as cycads, ferns and conifers, some of which are known from Triassic strata throughout the world. A Jurassic flora occurs also and this is comparable to that found in neighboring Egypt.

UNSTABLE SHELF

On the southeastern extremity of the Arabian platform there were episodes of emergence and also of profound subsidence, marked by the deposition of radiolarian cherts. In the Middle and Late Cretaceous massive conglomerates evidence a spectacular marginal collapse and contain clasts as young as the Albian overlying an unconformity which progressively removed all of the Early Cretaceous. During the Turonian to Early Maastrichtrian, Tethyan basinal and ophiolitic rocks of the Dibba zone were transported into place from the east-southeast and after their emplacement, mid-Tertiary compressional deformation produced large scale "whale-back" folds, in places overturned towards the west. Some of the folds are as much as 15 km in width and generally trend north-south. Thrusting has caused previously underlying Late Cretaceous tectonic units to override formerly overlying ones, thus reversing the stratigraphic succession. On the Hagab thrust, the complete shelf carbonate sequence of the Musandam Mountains has been thrust west-northwest over previously higher Hawasina and Haybi thrust sheets. The maximum extent of translation is here in excess of 5 km. This Tertiary folding and thrusting can be correlated in time and space with the Zagros fold belt of south-western Iran (3).

Early to Middle Cretaceous shallow marine carbonates of the northern Oman foredeep show a five-stage sequence of calcite sedimentation, fracturing and stylolitization, these stages correlating with the tectonic history of the northern Oman orogeny. Stylolite formation during rapid burial beneath abyssal Campanian fordeep sediments restricted hydrocarbon migration to Late Cenomanian and Early Campanian. In northern Iraq, Late Campanian to Maastrichtian sediments exceed 2.5 km in thickness and are well developed, representing three facies, namely, clastics at the top, reefal in the middle and open-marine at the bottom.

Radiolarian biostratigraphy demonstrates that Turonian beds overlie the Troodos ophiolite in Cyprus and are contemporaneous with similar strata overlying the Semail Ophiolite Belt of Oman. The Troodos ophiolite is one of several outcropping along a belt extending eastward from Cyprus across Syria, Turkey and Iran to Oman. This ophiolite belt is part of the Alpine mountain chain forming the northern border of the Arabian-African Plate. It is believed that the Troodos and the other ophiolites of the belt were formed in a series of short, spreading centers offset by transform faults above a subduction zone (4).

The Early Cretaceous of northwestern Saudi Arabia is marked by the influx of fresher water. Clastic limestones resembling those of the Jurassic were deposited in this and anhydrite deposition ceased. The rocks are very uniform and some of the calcarenites are very extensive. It is inferred that such colossal concentrations reflect a vast submarine platform just below sea-level resembling the existing Bahama Banks. Deposits of this age do not occur on the southern margin of the Arabian Peninsula, but this is more likely to be a consequence of truncation than non-deposition. Similar neritic carbonates are found in the center of the Rub al Khali, Qatar, Oman, coastal Iran, Kuwait and Central Iraq. The basal formation is chalky, aphanic limestone and it is followed by calcarenites and sandstones which are succeeded unconformably by Late Cretaceous sandstones and limestones. The sandstones were deposited roughly on the crest of the Hail Arc during an episode of emergence. Progressive inundation is evidenced by an upward change to shallow marine facies and later a final marine transgression which formed the Aruma Limestone.

The Hail Arc splits the Arabian Shelf into the Widyan Basin in the east made up of Cretaceous and younger sediments and the Tabuk Basin to the west including mostly Paleozoic rocks. In addition, at Khawr umm Waal, there is a trough composed of a set of grabens and synclines related to the Hail Arc which probably extends into southern Iraq to form the Rutbah High and, across Syria into Turkey creating the Mardin High. These highs may have had paleogeographic significance in constituting barriers in the Tethys Ocean (5).

In Jordan, there was renewed uplift commencing in the Kimeridgian and preceding the formation of Early Cretaceous sandstones which lie unconformably on various rocks. They are mostly marine in the northwest and were extended to the east of the present Jordan River by an Albian transgression. Continental sandstones formed elsewhere, gradually attenuating in the extreme southeast of Jordan. During the Cenomanian, marine calcareous sediments of a transgressive ocean blanketed the clastics with successively younger formations. By the Paleogene, the whole country was submerged as is evidenced by calcareous sediments covering thin clastics of the Late Cretaceous or even the Paleozoic sandstones in southeastern Jordan. Here, there is a well-preserved flora of this age recording the humid tropical climate which then prevailed. Vast amounts of silica from the hinterland were transported into the advancing Tethys and were fixed as widespread chert layers together with limestones and marls. Some limestones are bituminous reflecting euxinic conditions of basinal deposition and there are also abundant phosphorites and lumachelles of oysters in the Campanian, Maastrichtian and Paleocene.

The Mediterranean shelf of Israel is situated between the Arabian craton and the successor to the Tethys. It ranges in width from 3 to 25 km and is underlain by 4 to 9 km of Mesozoic and Tertiary strata, which, in part, represent a westward extension of the onshore post-Triassic sequence. Throughout its known geologic history from the post-Triassic until recent times, the existing shelf area straddled the transition zone from epicratonic sedimentation on the northwestern flank of the Arabian Shield to deeper water sedimentation in the ancestral Mediterranean (6).

In the Early Cretaceous, a Lebanese Trough was sinking. The downward movements were compensated for by fluviatile sedimentation intermixed with volcanic materials. Peat bogs grew from the plant debris of coniferous forests. The landmass was probably colonized by Araucaria and its predecessors since the Early Mesozoic. In the coal derived from the peat, Neocomian ambers are preserved in place, delta sediments containing reworked ones. The succession is 600 m thick at most and all its amber is derived from gymnospermous araucarias. There were occasional interruptions in the continental succession when inundations from the Tethys Ocean took place. Strata then deposited very rarely contain any of them. Near the end of the Neocomian, this sea again invaded and, for the first time, filled the trough. However, the low rate of subsidence caused an accumulation of coastal deltaic deposits which are full of amber. Most of them came from the Arabian-Nubian Shield, then lying southeast of Lebanon. The first major marine transgressions were probably linked with climatic changes to which the monkey puzzle trees seem to have reacted by increasing their resin production. It is interesting that, starting with the Early Aptian, marine conditions became more important in the region of the trough and caused an appreciable decrease in amber accumulation. From the Albian onward in the Levant, no more amber formed and this is ascribed either to extinction or migration of the araucarian forests.

It is clear that the ancient Egyptians knew many materials of which there is now no archeological evidence. This can be inferred from the state archives of Pharaoh Akhenaten at Tell el Amarna in which an assortment of minerals is listed as having been sent to him. Some of these are quite easily recognized, but others are not. Among the obscure ones are "hulalu" which may be white lead and "abasmu" which could be amber. Neither of these is known from Egypt, but, since they were both imported in the New Kingdom, they may well be mentioned in inscriptions. In the texts, attempts have been made to render the Akkadian "hulalu" into Egyptian, for instance by using the word "hrr". Both of these may be translated as "ceruse". From this it may be concluded that "abasmu" was also a foreign term. The modern "amber" is of Arabic origin, coming from "anbar", the French referring to it as "ambre". The Greek "elektron" also has this meaning, deriving from "elektor". This latter word means literally "shining" and refers to the effect of friction on amber.

The significance of the Lebanese amber is not its ornamental, but its scientific value. Formed in the Mesophytic Era during the dawning of angiosperms and mammals, it is the oldest fossil resin known to have trapped a large variety of insects. These include Orthoptera, Heteroptera, Homoptera, Coleoptera, Hymenoptera and Diptera and even aphids and feathers were preserved in it as well. Because of its age, the Levantine amber may shed light upon a number of fascinating enigmas. Among these are the actual appearances of some of the smaller terrestrial organisms living over 100 Ma ago and otherwise only found in fragmentary form as well as the type of insulation against cold, such as cocoons, hairs and feathers, which many of them employed. In addition, there is the matter of the paleoecologic niche enjoyed by Cretaceous animal taxa still living today on flowering plants. At that time, such a flora did not exist and the question arises as to whether they lived in conifers or in some other unidentified habitat. Finally, some creatures which now live on mammals may have left traces in ambers. If such can be found, an extraordinary modification of their living habits and an exciting adaptation happened.

Rudist fringing reefs are well developed in Albian to Turonian strata in Israel which attain a thickness of over 600 m. These are divided laterally into shallow shelf carbonate rocks accumulated on the Arabian craton and fine detrital carbonate deposits on its margins. The major rudistid belts developed during three depositional cycles and fringed the outer margins of the shelf platform. Their morphological characteristics show that the colonies developed into wave-resistant structures on a steep-sloping bottom facing the open sea.

Clastics and biogenics

BIOLITHIC ORES

A Late Cretaceous phosphate belt extends from Syria, Lebanon through Jordan and Israel into Egypt and thereafter North Africa, where it persists into the Eocene. The exact age of the deposits in the Syrian-Israeli region was a matter of dispute for many years. Libycoceras is restricted to the Tethys and the Late Campanian in Israel and Jordan, therefore not indicating a Maastrichtian age where it occurs. In fact, both the ammonite and the foraminiferal zonations clearly show that the Late Cretaceous high-grade phosphate deposits in Israel and adjacent countries are of Late Campanian and not Maastrichtian age. However, the phosphates comprise part of a lithostratigraphic unit of Campanian age of which the boundaries are diachronous so that they occur in Israel also in strata of Santonian to Early Campanian, Maastrichtian and occasionally Paleocene and Eocene formations as well.
In Syria, there are several phosphatic horizons of which the main exploitable ones are in the Late Campanian, Maastrichtian and upper Middle Eocene, although other phospate-bearing rocks occur in the Paleocene and Late Eocene. The Syrian phosphates are widely developed in the composite folded complexes of the Palmyra Mountains which are further complicated by extensive fracturing. Here, folded phosphate deposits are found in the Campanian which contains two major beds with calcium phosphate concentrations usually over 44%.
Cenomanian and Turonian rocks are well represented in the Galalas and are separated from each other by an ammonite bed at the base of the Turonian containing Pseudotissotia barjonai, Fagesia thevestensis, Hemitissotia cf. cazini and Leioniceras quassi. In the Gulf of Suez, the entire Cretaceous is represented by limestones and clastics.
During a major regression at the end of the Jurassic and in the Early Cretaceous, non-marine to paralic detrital deposits accumulated in six large basins between Algeria and the Arabian-Nubian Shield. The Ghadames, Sirte and northern Egyptian basins fringed the cratonic periphery of northeastern Africa. In the south, actually within the craton, occur the Murzuk, Kufra and southern Egyptian basins. Adequate data are available for the northern basins, but the reconstructions of the southern ones are merely suggested.
The regressive Nubian facies made up of low

sinuosity stream and fan delta deposits resulted from high detrital influx near the end of the Jurassic and in the Middle Cretaceous. The Ghadames, Murzuk and Kufra Basins became filled with several hundred meters of clastics and the three north-eastern basins filled with up to 2 km of detritus. A Tertiary flora comprising leaves, flowers and fruits from a fossiliferous claystone of the younger Nubian sandstones near Khartoum testifies that this formation is not entirely Cretaceous in age as had been thought, but ranges into the Early Neogene, at least in this region. The plant evidence shows that the paleoenvironment was a lacustrine one with adjacent tropical forests. In northern Egypt, the regressive sequence succeeded earlier Mesozoic marine sedimentation. In the Sirte and southern basins, correlative deposits accumulated on Precambrian and Paleozoic terrains after earlier Mesozoic uplift and erosion. As the detrital influx into southern Tunisia and Libya in the west and into Israel in the east diminished, an Albian to Early Cenomanian Tethyan transgression was initiated and, by the Late Cenomanian, had flooded the entire cratonic margin. It also spread south into the Murzuk and southern basins and also on to the Arabian-Nubian Shield. Possibly, the latest Jurassic, earliest, Middle and Late Cretaceous Tethyan transgressions resulted from worldwide sea-level rises coupled with regional tectonic control (7).

Faulting, uplift and erosion of pre-Cretaceous rocks of the Arabian-Nubian Shield initiated an episode of sedimentary accumulation. Thus, the sandstone sequences in the Natash Basin and on the surrounding Eastern Desert Platform comprise a lower, regressive, a middle marginal marine and an upper prograding paralic facies. The formation may correlate with lithofacies in the Dakhla Basin of central to southwest Egypt and along the lowlands fringing the Gulf of Suez. Except in Aptian times, an episode of regression prevailed through much of North Africa before a Cenomanian transgression marked a renewed extension of the Tethys Ocean. As the sea flooded southward, this was interrupted by a rapid northward progradation of the paralic facies supplied by uplift and erosion of the Arabian-Nubian Shield in Late Turonian-Coniacian time (8).

Paleomagnetic data from on the Cretaceous Nubian Sandstone and associated volcanics and iron ores in the Eastern Desert show that the climatic conditions at that time were tropical to subtropical. This accords with the position of the paleo-equator and

the evidence of included fossils. Cretaceous sandstones in Central Sudan also contain an oolitic iron ore of no economic importance, the first discovered south of Wadi Halfa. The paleoenvironment was continental, variably fluviatile, lacustrine or palludal, with occasional shallow marine episodes. In the Late Cretaceous, marine sediments occur with dominant calcareous platform deposits. During this time Egypt occupied a paleolatitude about 2,000 km south of the paleo-equator. Crustal movements went on during the accumulation of the sandstone and were accompanied by Late Cretaceous volcanicity in Wadi Natash. Paleomagnetic investigation corroborates that Africa has not undergone continental drift and there has been no related polar wandering during the period 210 to 85 Ma ago.
The Abu Ballas Formation is the first recorded Early Cretaceous marine transgression within the Nubian Group of southwestern Egypt, transgression and regression probably having occurred in the Aptian. Exposures are situated between the Gilf Kebir and the Abu Tartur Plateau. Above and below the formation there are fluviatile clastics.

Figure 5.3: The Early Cretaceous Abu Ballas Formation exposed in the southern part of the Western Desert, Egypt.

The Abu Ballas itself is mostly composed of red and green claystones with some intercalations of siltstones and fine grained sandstones (Figure 5.3). Its thickness averages 20 to 30 m and to the west, south and east, it peters out. Well-preserved plant remains were recorded together with an extensive fauna. This latter includes foraminiferids, solitary corals, many lingulids, abundant benthonic molluscs, arthropods, fish remains, ichnofossils and coprolites. Deposition was in shallow, rather brackish water (9).

During the Cretaceous and Early Tertiary, the southern part of the Western Desert was attached to a marginal cratonic depression in the north of the African continent. Alternating marine transgressions and regressions caused deposition of marine and continental sediments in the peripheral, increasingly shallowing parts of the depression. Although continuous subsidence is indicated by the deposits, local erosional unconformities demonstrate tectonic uplifting in the rims of the Upper Nile and Dakhla Basins since the Cenomanian. In the southern part of the Dakhla Basin and around the Kharga Oasis, there is complete preservation of the Cretaceous stratigraphic sequence. In the southern Upper Nile Basin, lack of Early and Middle Cretaceous sediments indicates an episode of pre-Campanian erosion. However, in the southern part of the Kharga uplift, which occasionally served as a western boundary between the Upper Nile and Dakhla Basins, pre-Campanian deposits are partially preserved. In the Cretaceous and Early Tertiary, a thick marine and continental sedimentary sequence accumulated throughout the entire area. Fluvial sediments accumulated in continental basins from pre-Aptian to perhaps the Campanian and, due to short transgressions, shallow sea deposition took place as well. In the Early Campanian, the sea returned and persisted until the Early Eocene (10).

Three marine deltaic cyclothems can be distinguished in the Nubian Group of the Dakhla Basin, attaining a thickness of 1 km. A complete cycle comprises basal transgressive delta-front sands containing ichnofossils, then fine clastics of the pro-delta to shallow off-shore area succeeded by sheet sands sometimes grading into marsh and estuarine deposits. The deltaic cyclothems are probably connected with minor onlaps preceding the Campanian transgression of the Tethys Ocean over Africa. There is an ongoing transition from the uppermost deltaic cycle to the marine facies of the Campanian (11).

Figure 5.4: Colossi of Pharaoh Rameses II hewn out of Cretaceous Nubian Sandstone at Abu Simbel.

The present day Kharga is referred to by Herodotus as the "City of Oasis" which was also called the "Island of the Blessed". He recorded a punitive expedition against the Ammonians who lived at the Siwa Oasis. The campaign was started in 525 B.C. by King Cambyses II, son of Cyrus the Great, with an army said to have included 50,000 men. Strangely, the operation began at Thebes rather than Memphis which is much nearer to Siwa. Probably, the guides responsible for leading the army wrongly imagined, as did Herodotus also, that the country of the Ammonians lay due west of Thebes. After departing from Kharga, no doubt on a westerly course, they disappeared in the vast sea of sand dunes which stretches continuously southeast from Siwa to beyond the latitude of Thebes. Alexander the Great in 331 B.C. knew the geography better, commencing his journey to the Oracle of Jupiter Ammon at Siwa from Paraetonium, the modern Matruh, and reaching it after an eight days march to the south.

The Kharga Oasis Depression is bounded on the east, north and west by escarpments. The stratigraphy of the exposures may be summarized. Overlying a crystalline basement is a variety of sedimentary formations. First is the pre-Maastrichtian 700 m thick Nubia facies composed of sandstones and shales (Figure 5.4). Above this is the fossiliferous Dakhla Shale (100 m thick) with an initial 7 m of phosphates, limestones and shales. Among important index fossils are Exogyra overwegi and Libycoceras berisensis from the Middle Maastrichtian succeeded by the Paleocene Balutchicardia beaumonti. Then follows the 15 m thick Tarawan Chalk, the Esna Shale (more than 100 m thick), the Thebes Formation, Early Eocene limestones with marls and shales contain-ing Nummulites deserti, and superficial loose deposits. Since the Early Maastrichtian, sedimentation was accompanied by epeirogenic movements causing a marked thinning of individual units on top of rising domes. The top of the folded Late Maastrichtian is marked by a widespread disconformity. In the Late Paleocene, a marine regression took place.

Late Campanian phosphatic sediments of Egypt start at the top of the Nubian Group and reach a maximum along a central latitudinal belt which extends, with interruption from the New Valley, across the Nile Valley to the Red Sea. The Late Cretaceous Duwi Formation occurs on the Red Sea coast. Palinspastic lithofacies reconstructions demonstrate lateral and vertical facies variations. Phosphorite is present

Figure 5.5: Late Cretaceous phosphates at Abu Tartur in the Western Desert.

as pelletal-bearing beds often containing skeletal debris and usually bound by either carbonate or siliceous cement. One proposal for the formation of the bedded phosphorites involves precipitation of calcium phosphate compounds (collophane) from interstitial waters within anoxic sediments and subsequent reconcentration by physical processes of winnowing and reworking (Figure 5.5).

Total phosphate reserves are more than a billion tons. To the west, glauconitic sands and siltstones dominate, whereas in the east, limestones, cherts and bituminous shales are recorded. Clastic phosphatic particles comprise bone fragments, teeth, peloids, coprolites and lithoclasts. Distribution, size and shape of particles suggest accumulation of phosphates under well-aerated and agitated water conditions. The major phosphate phase is francolite and carbonates (mainly dolomite), quartz, clay minerals, sulfides and organic matter are present as well, their chemical composition being marked by enrichment of elements such as cobalt, nickel, vanadium and zinc being precipitated with organic matter under more or less anaerobic conditions. At the Abu Tartur Plateau, 50 km west of the city of Kharga, a large phosphorite deposit has been found in terrigenous-carbonate Maastrichtian sediments, the seam being 4 m thick and containing an average of 25.5% phosphorus pentoxide. The reserves are calculated at a billion tons (12).

Ammonites ascribed to the Bostrychoceras polyplocum zone disclosed that the Western Desert phosphates are Late Campanian in age and correlate with equivalents in the Nile Valley and along the Red Sea. Fish, mainly sharks, teeth are abundant and so are reptile remains, mostly of turtles and crocodiles. An uplift in Bahariya at the Campanian-Maastrichtian boundary is manifested by a distinct angular unconformity between the phosphates and the overlying chalk. Contemporaneous uplift occurred at Kharga (13).

Iron ore deposits cap the Gebel Ghorabi in the Bahariya Oasis and are of Middle Eocene age. The ore comprises pisolites and red and yellow ochre. It is mainly made up of goethite, with variable amounts of hematite, magnetite and pyrolusite. It may derive from the hydrothermal activities of the Tertiary volcanicity which is manifested by flows, sills and dikes in the area. Part of it may well have formed during a lagoonal episode by slow weathering and leaching of ferruginous sandstones in the center and south of the oasis. Transported material may have precipitated directly into the lagoon or permeated underlying rocks, replacing them selectively. The ore bodies comprise a major supply source for the Helwan steelworks, replacing the formerly used, but inadequate deposits in Aswan.

There is a perennial controversy as to the antiquity of the employment of iron in Egypt. The word "bi" may have been used both for this metal and meteoric material. Interestingly the Coptic word is "benine", translated as iron. However, it seems clear that "bi" has a very limited application. Where utilized in reference to ropes by which the ascent to heaven is made or to the heavenly throne, probably meteorites were meant. In fact, some believed that metallic iron was first obtained from meteorites. In late religious texts, the word is used in reference to weapons and there is no doubt the the red and yellow ochres were valuable pigments for paintings and personal cosmetics. The connection of "bi" with the heavenly vault does not rule out its application to iron ores because the Egyptian concept of meteoric material was far from exact. Actually, until the New Kingdom, there were no precise terms for hematite and magnetite, terms for these first appearing in the Book of the Dead, "bi ksy" for the former and "bi sm" for the latter which perhaps also described red brick. All iron ores were universally confused in antiquity and most primitive peoples thought such materials to be of supernatural origin, perhaps connected with lightning. In sum, the meanings of "bi" cannot be fully established, but it undoubtedly comprised substances, including iron ores, which the old Egyptians took to be of meteoric origin.

On the western side of the Murzuk Basin, 800 m of mostly fluviatile, reddish sandstones with intercalated clays are exposed and overlie marine Carboniferous beds. The sequence was divided into basal Tilemsin Formation and overlying Messak Sandstone which were assigned to the Permo-Trias to Early Cretaceous time interval. The Tilemsin derived from an area of mild weathering under semi-arid to semi-humid-tropical climate. The Messak Sandstone is predominantly kaolinitic and its sediments represent reworked products of intense lateritic weathering under humid-tropical conditions. However, the transitions are gradual from which gradual climatic changes are inferred. The Messak Sandstone belongs to the Nubian facies characterized by the products of humid to tropical weathering.

The first trans-Saharan transgression commenced in the Late Cenomanian and reached its maximum in the Early Turonian. Entering through subsiding areas in North Africa the sea spilt on to the stable craton. Around Middle Turonian time, the Central Sahara was land and the sea, which had persisted in the central part of the Benue rift, extended again during a minor Coniacian inundation which ended in central and eastern Niger. Tethys and South Atlantic may have been joined during the Cenomanian-Turonian. A third, extremely extensive epicontinental submergence began in the latest Campanian and reached a climax in the Early Maastrichtian. At least in the Central Sahara, there was a retreat of the sea in the later Maastrichtian. A final engulfment reached its zenith during the Paleocene and thereafter the sea permanently withdrew from the northwestern African hinterland (14,15).

REFERENCES

1. SHARIEF, F.A., 1981. Transgressive-regressive intervals of the Tethys Sea in the Middle Triassic Jilh Formation, Saudi Arabia. Newsl. Stratigr., 10, 3, 127-139.

2. DRUCKMAN, Y., HIRSCH, F. and WEISSBROD, T., 1982. The Triassic of the southern margin of the Tethys in the Levant and its correlation across the Jordan rift valley. Geol. Rdsch., 71, 919-936.

3. SEARLE, M.P., JAMES, N.P., CALON, T.J. and SMEWING, J.D., 1983. Sedimentological and structural evolution of the Arabian continental margin in the Musandam Mountains and Dibba zone, United Arab Emirates. GSA, Bull., 94, 1381-1400.

4. BLOME, C.D., IRWIN, W.P., 1985. Equivalent radiolarian ages from ophiolitic terranes of Cyprus and Oman. Geology, 13, 401-404.

5. SHARIEF, F.A., ROGERS, J.J.W., 1980. Sedimentary facies and regional significance of the Sakaka Sandstone, northwestern Saudi Arabia. Newsl. Stratigr., 8, 3, 171-179.

6. PRASAD, G., LEJAL-NICOL, A. and VAUDOIS-MIEJA, N., 1986. A Tertiary age for Upper Nubian Sandstone Formation, Central Sudan. AAPG, Bull.,70, 2, 138-142.

7. VAN HOUTEN, F.B., 1980. Latest Jurassic-Early Cretaceous regressive facies, northeast Africa craton. AAPG, Bull., 64, 857-867.

8. Van HOUTEN, F.B., BHATTACHARYYA, D.P. and MANSOUR, S.E.I., 1984. Cretaceous Nubia Formation and correlated deposits, eastern Egypt: Major regressive-transgressive complex. GSA, Bull.,95, 397-405.

9. BÖTTCHER, R., 1982. Die Abu Ballas Formation (Lingula Shale) (Apt?) der Nubischen Gruppe Suedwest-Aegyptens. Berliner geowiss. Abh., A, 39, 1-145.

10. HENDRIKS, F., KALLENBACH, H., 1986. The offshore to backshore environments of the Abu Ballas Formation of the SE Dakhla Basin (Western Desert, Egypt). Geol. Rdsch., 75, 445-460.

11. FAY, M., HERRMANN-DEGEN, W., 1984. Mineralogy of Campanian/Maastrichtian sand deposits and a model of basin development for the Nubian Group of the Dakhla Basin (Southwest Egypt). Berliner

geowiss. Abh., A, 50, 99-115.

12. GERMANN, K., BOCK, W.-D. and SCHRÖTER, T., 1984. Facies development of Cretaceous phosphorites in Egypt: Sedimentological and geochemical aspects. Berliner geowiss. Abh., 50, 345-361.

13. DOMINIK, W. and SCHAAL, S., 1984. Notes on the stratigraphy of the Upper Cretaceous phosphates (Campanian) of the Western Desert, Egypt. Berliner geowiss. Abh., A, 50, 153-175.

14. REYMENT, R.A., 1980. Biogeography of the Saharan Cretaceous and Paleocene epicontinental transgressions. Cret. Res., 1, 299-327.

15. KOGBE, C.A., 1980. The Trans-Saharan Seaway during the Cretaceous. In: The Geology of Libya, 1, ed. SALEM, M.J., BUSREWIL, M.T., 91-96, Acad. Press, London, UK.

Chapter III-6

THE SYRIAN ARC EMERGES

"Thou shalt be visited of the Lord of Hosts with thunder, and with earthquake, and great noise, with storm and tempest, and the flame of devouring fire". Isaiah, 29, 6.

TOLLING OF THE KNELL

Until recently, the Cretaceous-Tertiary boundary, distinguished by a global unconformity recorded in the hebetic Syrian Arc, was regarded as extremely difficult to trace, despite the apparent continuity of sedimentation between the two. However, the Maastrichtian may be clearly differentiated from the Paleocene because it is defined by the extinction of dinosaurs and other reptiles, ammonites, belemnites, rudists, inoceramids and globotruncanids. Unfortunately, these do not vanish at any one stage of the Late Cretaceous, rather fading away at differing times in various facies. The disappearance of so many species and genera of the animal kingdom could have several explanations. Among these are widespread paleogeographic changes through tectonism or drastic environmental modifications induced by climatic alterations.

Except in parts of Egypt, an extensive break marks the boundary between the eras all over the Levant and may be due to uplifts. Recently, an extraterrestrial cause has been postulated for the Cretaceous-Tertiary extinction event, said to be the last of several such episodes taking place in the Phanerozoic Eon. At this time, iridium increases of orders of magnitude were recognized in marine sediments from several places around the world. It is speculated that this phenomenon is the result of an impact of a large asteroid with a diameter of

approximately 10 km at an unknown place (1).
This would have been accompanied by a huge atmospheric dust cloud, similar to, but unimaginably greater than, that emitted by Krakatoa in the last century, with a concomitant arresting of photosynthesis and a disruption in the food cycles. There is no fossil plant evidence in favor of the idea. Moreover, no crater resulting from the event has been found on land. Consequently, an oceanic site has been sought, but with conspicuous lack of success to date, although the Amirante Basin roughly 1,000 km east of Kenya has been proposed. This has a subcircular ringed appearance and it is claimed that there may be possible evidence in East African continental margin deposits around the Somali Basin for expected earthquake wave and tsunami effects related to this hypothetical clash. Of course, the Amirante Trench and arc system can be explained as a volcanic arc. Actually, oceanic tholeiitic basalts have been dredged from it and dated at 82 Ma by K-Ar. As the trench cuts the spreading fabric in the Amirante Passage, the objection has been made that, if an Early Tertiary northward ridge jump occurred there, it is anomalous that the purportedly younger Amirante volcanic arc basalt should have given an age exceeding 65 Ma. This is difficult to uphold because of the error margin involved in this type of dating. In the present case, it amounts to 32 Ma and if the negative half is deducted, a figure of 66 Ma is arrived at and practically coincides with the 65 Ma of the caveat. The northward ridge jump alluded to above is alleged to be manifested by a cutting of the African-Antarctic boundary across older crust which arose at the junction between the African and Indian Plates. Separation of the Seychelles Bank from India by a more than 500 km ridge jump also attests to reorganisations of western Indian Ocean tectonics taken to relate to the impact. All this is rather speculative, but perhaps less so than estimates of the kinetic energy hypothesized for the event (2,3).
At the Cretaceous-Tertiary, or, as it is now modish to call it, the K/T boundary, the clays contain microtektites of disputed origin. Incidentally, these have also been collected from the Oligocene of deep sea cores, thus removing any special significance from them as stratigraphic boundary markers or indicators of mass extinctions. Some assign them to an oceanic source and others regard them as terrigenous. They may be diagenetically altered impact droplets of basaltic composition. A

consensus seems to be emerging around an oceanic impact, but shocked quartz grains have been found in Montana and imply a continental target. The problem could be resolved if an oceanic impact took place near the margin of a continent. If this was active and convergent, an advantage would accrue for modern catastrophists since the impact site would have conveniently disappeared. Hence, it cannot be detected even in the oceans! Anyhow, dynamic deformation producing shock features in minerals from rocks of this same age can easily result from internal processes and certainly do not require any impacts of extraterrestrial objects.

"THOSE MOST THEOLOGIC HILLS"

This quotation from R.B. Cunningham Graham may be aptly applied to the whole of Arabia Petraea, although he intended it only for Lebanon. In this focal point of cultural and religious history, traversed by invading armies throughout the ages, a suordinate of the Old Man of the Mountain in Persia constructed his headquarters in the time of the Crusades. Marco Polo relates that he was visited by Count Henry of Champagne who noticed two young men in white sitting on a tower of his castle. The Ismaili Imam questioned whether the Count had such devoted followers as he and, without waiting for a response, made a sign to the two whereupon they jumped from the tower to their deaths.
On precisely which mountain this ominous castle existed, if indeed it is not legendary, cannot be determined because most of the country is a plateau with many peaks of the Lebanon range, running for about 170 km. The elevations range from 715 m in the south near the Qalat esh-Shakif, a Crusader castle, to well over 2,000 m at Gebel Baruk. The plateau comprises a broad and gently folded anticline. At the base of the sequence are up to 400 m of Jurassic limestones overlain by Early Cretaceous sandstones making up an important aquifer. The thickest beds are limestones, which amount to 1,000 m, ranging from the Cenomanian into the Turonian. Chalky and flinty beds complete the sequence, except in the south, east and northwest where up to 50 m of Eocene nummulitic chalk is preserved. Near Beirut, there are also Miocene limestones which are rich in corals, echinoids, pectinids and oysters. There were two volcanic episodes, one in the Early Cretaceous

and the second in the Plio-Pleistocene.

Toward Syria, the Bekaa Valley, a northernmost extension of the Aqaba-Dead Sea Rift Zone, separates these mountains from their easterly neighbors, the Anti-Lebanon. Pivotal to the latter and the Palmya Chain is the often snow-covered Mount Hermon, a limestone massif 2,760 m high with a SW-NE Syrian Arc trend. Structurally it is a huge faulted, fractured and partly collapsed dome composed of Late Jurassic limestones, arenaceous Early Cretaceous and succeeding limestones which followed the Cenomanian Transgression. Volcanic rocks also occur.

Referring to the Anti-Lebanon Mountains rising to a plateau 1,800 m high , these were originally thought of as an eastern wing of a dual system including the Lebanon Mountains. However, it has long been recognized that they are almost mirror images of each other with similar sequences and structures in both.

A mountain chain separates northeast from Gebel Qasyoun, a marginal overthrust fold of the Ghouta Oasis in Damascus, diverges from the Anti-Lebanon. The Qasyoun is one in a series of anticlines which crosses Central Syria toward Tadmour, the old Palmyra, the gateway between the Sahl et Daou in the nortwest and the great Syrian Desert in the southeast. At Palmyra, the geological sequence comprises Late Cretaceous marls and dolomites of 200 to 300 m thickness overlain in adjacent areas by Paleogene limestones and Neogene conglomerates up to 300 m in thickness. The Turonian dolomite was quarried to supply the building stones for the ancient city of Palmyra. Diapiric anhydrite of indeterminate age underlies the whole sequence and produced two anticlinal structures at Palmyra. A couple of thermal sulfurous springs, one of them feeding an antique bath which is still in use today, deliver 150 and 20 l/sec, respectively, excess water going south to the enormous sabkha called Sabkhet Mouh.

South of the Lebanese and Syrian mountain chains, the Judean Desert in Israel is bounded on the west by an anticlinorial mountain chain and on the east by the Dead Sea Rift Valley. Folding is complicated by Senonian to Miocene NE-SW flexures and fissures. The Canaan Valley fault zone is almost parallel and 10 km to the west of the Dead Sea comprising a lineament more than 100 km long. A fault complex near Mount Holed forms a long narrow graben, once part of a dome which developed continuously during Santonian-Campanian times. This correlates with structures in the northern Negev comprising part of the Syrian Arc in which folding started during the Senonian. Jointing started with the formation of two monoclines during the only known regional tectonic activity of the Late Cretaceous. There is an accordance with joints developed in these during the Laramide Orogeny and they differ from those within non-folded Eocene rocks (4).

Figure 6.1: Structures of the Syrian Arc.

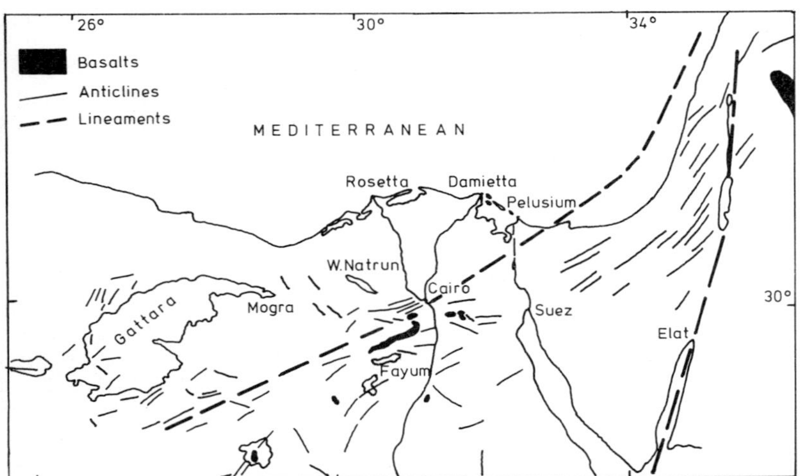

An impressive bastion of the Syrian Arc is found in the deserts of northern Sinai (Figure 6.1). This is the 735 m high, wadi-dissected Maghara Massif which is 55 by 30 km in extent. It comprises two NE-SW doubly-plunging anticlinal features. The subsidiary one is disturbed by a shallow syncline. Interestingly, not a syncline, but a fault trending northeast and with a throw of more than 800 m, separates them. In addition, there is a swarm of other faults in the area, the northwesterly ones being mostly of the hinge type with downthrows of 5 to 100 m. Probably, there has been more than one episode of pre-Eocene faulting since northeasterly fractures appear to be younger.

The northernmost part of the Gulf of Suez marks the intersection of the two major tectonic trends in northeastern Egypt, which are the Clysmic faults and the Syrian Arc folds. Its central part contains the strongly asymmetric Miocene to Recent Abu Darag Basin which has more than 6 km of sediments in it. The southern part of the area embraces the down-faulted offshore eastern extension of the large

Wadi Araba structure. As the result of the intersection of these last two widely differently elevated features, the central basin was displaced to the eastern side of the Gulf. The Wadi Araba runs northeast to southwest inland from the Gulf of Suez and comprises an inlier with a rim of Nubian sandstones enclosing a core of Carboniferous clastics and carbonates. The limestones and marls are exposed near the bottom of the surface section. In appearance, it resembles a heavily eroded, fault-bounded and extended anticline, a textbook example of a geomorphic inversion. The northern Gulf of Suez shows a northerly thickening of the pre-Miocene section and a more marine facies. Jurassic sediments occur with a probable maximum development in western Central Sinai.

AN ERUCTATING CATENA

As was previously indicated, the powerful global Laramide tectonism affected Mesozoic to Early Tertiary strata in the Levant, forming the Syrian Arc. When the African Plate was moving north, its folds and the complementary NW-SE and NE-SW fault systems together with a system of shear fractures originated from compressive stress which also acted on the unstable shelf of Egypt. Additional modifications in the strain pattern were produced by subsequent counterclockwise rotation of the Arabian Shield. It extends in a southwesterly direction and follows a sinuous path from the Palmyra Mountains through those of Lebanon to Gebel Maghara in the northern Sinai, along the Cairo-Suez Road to Abu Roash and the Oases of Bahariya and Farafra in the Western Desert of Egypt. The Arc in Egypt is sometimes divided into three trends, i.e. the Abu Roash, which includes the Shabrawet Hills adjacent to the Great Bitter Lake, the Halal which passes through Bahariya and the Wadi Araba traversing the Farafra Oasis. Between Abu Roash and Halal lies the Cairo-Suez district, which suffered folding in post-Jurassic, Nevadian, time. However, it was affected again during the Laramide Orogeny.

At the end of the Cretaceous and in the Early Paleogene, the arc was subaerial. Later, a marine incursion submerged it with limestone deposits. It emerged again at the close of the Middle Eocene. During this entire epoch, the sea was usually in retreat and when it terminated, continental conditions prevailed over the area. Deep-water deposits of the Late Cretaceous accumulated in the basins separating structural highs of the Syrian Arc. These have been affected by subsequent faulting of NW-SE and latitudinal Alpine orientation. The arc grew intermittently until it was completed in the Late Oligocene. Then, the Proto-Red Sea embayment was upwarped, thereafter to subside into the graben in which sea-floor spreading ultimately developed.

Erosion of Eocene deposits in the present Red Sea area attended a marine regression following elevation of land to the south. Accompanying this, an elongated dome with a northwestto southeast axis arose in a position almost coinciding with the existing Gulf of Suez. This collapsed into the graben of the Red Sea due to instability induced by the widespread Oligocene Alpine Orogeny. By the Miocene, the Cairo-Suez area was partly flooded although some islands of older rock remained. In the middle of this epoch, evaporites began precipitating in the lagoons of the newly formed Gulf of Suez.

The close of the Miocene witnessed yet another upwarping with continental emergence and this persisted into the Pliocene. It was responsible for the creation of the Tethyan and Clysmic fault trends in the Cairo-Suez region. Forty kilometers northeast of Cairo, the Gebel Um Qamar and Gebel El Hamza stand out as landmarks on the plain. They are the northern and southern flanks of a plunging asymmetric anticline of the Syrian Arc. The southern limb dips more steeply than the northern one, reflecting the northward movemnent of the African Plate and its incorporated Arabian-Nubian Shield. The oldest rocks are Oligocene clastics and overlying weathered basalt mounds succeeded unconformably by marine Miocene carbonates grading up into non-marine beds attaining a total thickness of around 100 m. The overlying Pliocene limestones are only 0.5 m thick and followed by more than 5 m of eolian sands, boulders, pebbles and lag deposits.

About 70 km from Cairo and east of Gebel El Hamza, there is another upwarp. This, the Gebel Um Raqm, rises 243 m above sea-level and is an extremely shallow anticline bearing from northeast to southwest over an area of 90 km². The sequence includes 30 m of Late Eocene fossiliferous limestones and dolomites followed by 35 m of Oligocene sands and gravels. Above, there are 60 m of fossiliferous, initially marine clastics and carbonates. Finally, there occur Miocene non-marine sands and gravels of 20 m thickness.

Figure 6.2: Diapiric fold in Cenomanian marls and limestones with adjacent faults. Abu Roash, Cairo.

In the Eocene limestone plateau bordering the cultivated Nile Delta west of Cairo, there is the hilly inlier of Abu Roash comprising Laramide folded and faulted Late Cretaceous clastics, marls, limestones, dolomites and chalk reaching a thickness of 250 m. The summit of these hills at Gaa is 150 m high. It is the site of the northernmost pyramid, that of the 4th Dynasty Pharaoh Dedefre, the son of the Pharaoh Cheops who died 2528 B.C. after a twenty three year reign. During this, he constructed his colossal tomb at Gizeh, one of the Seven Wonders of the World. It can be seen from Gaa together with the other lesser pyramids comprising a unique and unforgettable panorama including Sakkara and Memphis. After the death of his father, the new Pharaoh entombed the royal funeral barge before the south wall of the Great Pyramid. He placed his cartouche on the limestone blocks which hid it from the light of day until 1954. Dedefre was the first to introduce the name of the solar deity, Re, into his own, taking also the official title "Son of Re". He reigned until 2520 B.C., perhaps not long enough to complete his own tomb, but finding sufficient time to face it with red granite from Aswan.

At Abu Roash, the type locality for many fossils and

the standard section for the Late Cretaceous of Egypt, the first oil exploration well was drilled here and proved to be unproductive. Marine Jurassic was recorded from it. The surface of the limestones is an excellent seismic reflection layer utilized in the Western Desert in the continuing search for energy sources. Below the carbonate sequence, brown sandstones with few marine fossils belong to the Nubian facies and are assigned to the Cenomanian. The overlying Turonian contains Tethyan fossils such as rudists and actaeonellids succeeded by Senonian marls, limestones and chalk. In the latter, the Campanian ammonite Libycoceras is found. The Late Mokattam Formation of the Middle Eocene rests unconformably on the preceding folded Cretaceous and is covered by Oligocene gravels and basaltic lava sheets (Figure 6.2).

Structurally, the Abu Roash area comprises a number of broad anticlines separated by narrow synclines, the structural axes being directed southwesterly from the Nile Delta. The southeasternmost antiform extends as far as the Gizeh Pyramid Plateau, which is itself a marginal part of the Pliocene coastline. The major effects of the Laramide Orogeny occurred in the region during the Paleocene when the above-mentioned structures were formed. Concomitantly with the main rifting of the Red Sea, another deformational episode took place and resulted in an uplift of the basement and gentle folding of the Eocene limestones, including those at Gizeh. Subsequently, extensive faulting occurred, a major one crossing the Wadi El Qarn from east to west and perhaps constituting the Pelusium Line, with accompanying extrusion of basalt. Other faults fractured the structures, particularly to the north.

The Bahariya Oasis is one of the seven major depressions in the Western Desert of Egypt. Situated about 320 km southwest of Cairo, it has an oval shape extending 100 km from the northeast to the southwest and a maximum width of 45 km. It is totally enclosed by steeply sloping scarps which are dissected by wadis ending on its floor. The surrounding plateaus rise to a maximum of 300 m elevation above sea-level. The regional tectonics comprise the Ghorabi and El Heiz anticlines making up the Bahariya structural high with a northeasterly trend. The former is 30 km long and plunges to the north while the latter is twice as long and plunges in the opposite direction. There are three sets of major faults trending from ENE to WSW. More localized tectonic features are concentrated in the

northern part of the depression, especially along the scarps. The dominating structure is the Hafhuf syncline on the depression floor, lying between the anticlines (5).
The Syrian Arc is known to extend beyond the Bahariya Oasis to Farafra and Kharga. The structure of the latter is dominated by a main central NE-SW doubly plunging anticline, a faulted syncline, a minor anticline and a horst. The relevant beds include overlying Late Cretaceous chalk and compact limestone with Nubian clastics underneath, extending down 2,400 m to the surface of the Precambrian. Faults are difficult to see because of the vast blanket of wind-blown sand. However, there are two sets between the northeastern part of the depression and the Bahariya Oasis to the north, these being meridional. Structurally, the Kharga Oasis comprises a group of doubly plunging anticlines arranged en échelon mostly north to south together with many faults. Some parallel ones define a graben separating a couple of horsts. Two tectonic events affected the area. There was a pre-Maastrichtian episode which formed the Kharga and Beris Basins. Later, a post-Early Eocene movement produced the predominant folding and faulting in the region.
Subterranean structural highs occur in Egypt, a good example being the one at Matruh, their significance relating to their oil potential. Actually, most of the anticipated petroleum either did not form or escaped from these traps. The upwarps alternate with structural lows such as that underlying the Fayum. Apart from the Abu Roash deformation center, other buried NE-SW arc folds are near Burg El Arab and Moghra. Deep wells were drilled to investigate oil potential in such anticlines and demonstrated sedimentational hiatuses between the Mesozoic and the Cenozoic. The folds of the Syrian Arc may be still active, in which case geomorphic depressions like those of Moghra, Abu Mina and the eastern side of the Wadi El Farigh comprise "echos" of underlying anticlinal warps or folds. As well as the arc system, the northern part of the Western Desert contains the NW-SE Clysmic one. A prominent fold element in this latter is the Wadi Natrun anticline composed of heavily faulted Pliocene beds and another is the Wadi El Farigh anticline.
In Libya, marine Late Cretaceous and Tertiary sediments outcropping at Gebel Akhdar were moderately folded and faulted during the Santonian and Ypresian Ages 85 to 55 Ma ago. The area is a gently raised plateau composed of Late Cretaceous and Ter-

tiary sediments, mostly limestones, subordinate dolomites and marls. These were deposited at the southern margin of the Tethys Ocean. In Late Campanian or Early Maastrichtian times, a major tectonic event commenced with weak movements and reached its maximum intensity at the end of the Paleocene. When this orogenic episode took place in Cyrenaica, much of the area was uplifted, later folded and severely eroded. Less severe orogenic movements went on until the Ypresian when a transgression began, seas advancing into the basin from the north to lap on to the steep flanks of the folded highlands. The whole Gebel Akhdar area was again inundated during the Lutetian which was an age of tectonic quiescence. From the Middle Eocene to the Middle Miocene, the region was subjected to slight warping followed by oscillatory transgressions of a shallow sea. The youngest tectonic movements caused a gentle doming in the area associated with the downfaulting of certain zones. The Middle Miocene structural relief is retained. The main structural alignment is that of the Late Cretaceous folding running northeast to southwest and crossed by NW-SE and latitudinally striking faults. It developed in a marginal part of the North African Shelf. The Late Cretaceous and Eocene sediments of the present-day coastal area may have been deposited partly on the continental slope. Gebel Akhdar is the most important part of an extensive tectonic paleobarrier between the Tertiary epicontinental basin of northeastern Libya and the Tethys deep sea basin (6).

VOMITION OF MAGMA

Egyptian alkaline ring complexes were emplaced during three magmatic events in the Phanerozoic, in the latest of which they became associated with the development of the Syrian Arc. As is usually the case, William Shakespeare has some appropriate lines for such a scenario in Henry IV where he describes what happens when "Diseased nature oftentimes breaks forth in strange eruptions: oft the teeming earth is with a kind of colic pinched and vexed by the imprisoning of unruly wind within her womb; which, for enlargement striving, shakes the old beldam earth and topples down steeples and moss-grown towers".
Alkalic rocks are found in the south Eastern Desert of Egypt as ring complexes, ring and linear dikes,

lava fields and agglomeratic plugs. They are characterized by milky alkalic and exclusively sodic nature. Most ring complexes show a deficiency in silica, alumina or both. Trace element distribution patterns indicate a "within plate" non-orogenic setting for Wadi Natash Volcanics. The alkaline ring complexes in the basement between the Nile and the Red Sea range in diameter from 1 to 8 km and comprise several intrusive phases with plutonic rocks varying from gabbro to nepheline syenite, related volcanics such as rhyolites, trachytes and phonolites, numerous faults cutting both them and surrounding country rocks.

Potassium-argon ages for nine of the ring complexes in the southern part of the Eastern Desert range from Cambrian to Late Cretaceous and do not show any geographic pattern. The various units of the complexes provide ages identical within analytical uncertainties and this restricts the time available to emplace any one of them to only a few million years. They occur at shallow levels in the upper crust and are associated with contemporaneous volcanics and dikes. From Rb-Sr isochron and conventional ages together with K-Ar dates, three phases of igneous activity may be inferred. Initial strontium isotope ratios vary between 0.703 and 0.704 in the undersaturated complexes of El Kahfa and Mishbeh and the Wadi Natash Volcanics and a high of 0.704-0.706 in the more saturated masses of Silai and Bir Um Hebal. This reflects evolution by differentiation from mantle-derived basaltic parent magma for the first group and a similar source, but with limited crustal contamination, for the second group.

These Paleozoic complexes have no direct relationship with the modern Red Sea spreading events, although they may be correlated with large-scale fracture patterns. The 230 Ma phase of the Late Permian was marked by emplacement of the more saturated massifs of Gebel Silai, Bir Um Hebal and Zargat Naam. Its age may accord with those of the Late Paleozoic ring complexes in Nigeria, Niger and Cameroon. The Zargat Naam age of 247 Ma is regarded as a minimum age for the Nubian sandstones in its area.

The Paleozoic structure of Gebel Um Risha measures about 12 km from east to west and is partly covered with Nubian sandstones. There is an outer ring of granosyenite enclosing a mass of alkaline syenite and nordmarkite. Remnants of the old volcanic cone include trachybasalts and pyroclastics and are encountered in the central and western parts of the complex where they are partly affected by the syenites to form a variety of hybrid rocks. The structure is dissected by several northwest and northeast trending faults, many of which are concealed beneath the Nubian sandstones (7).

The Mesozoic complexes may be related both in pattern and in age to early rifting of the Red Sea, but they predate the actual sea-floor spreading by at least 80 Ma. The 140 Ma phase of the Early Cretaceous witnessed the emplacement of the more alkalic ring complexes of Mishbeh, El Naga, Nigrub El Tahtani and El Mansouri. The 90 Ma phase of the Late Cretaceous was seismically active and the alkaline volcanics of Wadi Natash were erupted 104 Ma B.P. together with the intrusion of the ring complexes of Gebel Abu Khruq 84 Ma B.P. and El Kahfa 81 Ma B.P. A detailed study of the former was made after the construction of the High Dam at Aswan. Eleven others were subsequently found in the southern Eastern Desert. However, it was found to be technically impracticable to produce alumina from any of the nepheline syenites, thus disappointing yet another economic development scheme. Some K-Ar ages extend the volcanic activity in Wadi Natash to 70 Ma ago.

The undated Gebel Hadyib ring complex is intruded into volcanics and pink, late orogenic granites, is oval in outline and measures 4 km meridionally and 1.5 to 3.5 km in width. It includes two superimposed ring structures, a southern, smaller one which is younger than a bigger northern one. The southern ring comprises an outer ridge and a conical stock separated by a ring wadi, the ridge and the stock being made up of alkaline syenite and nordmarkite. The northern structure is composed of successive incomplete ring ridges of the same rocks partly covered by uplifted remnants of the volcanic cone such as trachybasalts and pyroclastics. On the north and south sides, the complex is cut by many radial and oblique dikes of the solvsbergite grorudite series as well as by some pegmatite veins of syenitic composition. A number of NNW and ENE trending faults cuts the structure (8,9).

Of particular interest is the occurrence of peridot and nickel deposits in ultramafic rocks intruding the Cretaceous of St. John Island in the Red Sea about 60 km eastsoutheast of Berenice on Foul Bay, although only the former may be considered commercially viable. In fact, peridot has been exploited since Dynastic times, when cobras were employed to dissuade thieves and slaves did the mining. Rocks on the 4.5 km^2 large island include not only silica-deficient ultrabasics, but also sedimentary and metamorphosed sedimentary beds. These latter probably originated before rifting began in the Red Sea, later separating to form a partially submerged promontory. Subsequently, the igneous rocks came up from the rift zone and intruded and altered them. Metamorphics comprise granulites, schists and slates. The fact that the igneous rocks are extremely basic was critical in the formation of peridote in the first place. The main intrusives are peridotite and dunite which are composed not only of olivine but also locally contain spinel and biotite as well as the usual assemblage of minerals associated with such rocks. Peridots were and are mostly hand-dug from veins or veinlets in serpentinized peridotites. Specimens from this, the "Serpent Isle", have been found in archeological excavations in Alexandria. Crystals of forsterite, white olivine, several centimeters long may be found in the Smithsonian as well as in other reputable collections. The Romans called the island Topazos from which the ancient name for peridot, topazion, was derived. Such gemstones adorned the lost pectoral of the High Priest of Jerusalem, the King of Tyre and Queen Berenice, the wife of King Ptolemy I of Egypt. The location of the island was a state secret, possibly for millenia, but occasionally it was seized by pirates. Curiously, until it was rediscovered around 1900, the origin of the peridot used by the ancients was completely unknown. European peridots were often labelled as deriving from Egyptian sources, but were sometimes ascribed to the Levant in general and called Levantine peridots or evening emeralds. By 1906, the position of the island became public knowledge and mining on it was the monopoly of the Khedive until the Great War. After this, exploitation went on under government auspices until World War II. After 1958, the mines were nationalized with the result that operations collapsed and production ceased. However, illicit peridots continue to appear on the market even today and fetch prices of several hundred US dollars per carat (10).

REFERENCES

1. ALVAREZ, L.W., 1983. Experimental evidence that

an asteroid impact led to the extinction of many species 65 million years ago. Proc. Natl. Acad. Sci. USA, 80, 627-642.

2. HARTNADY, C.J.H., 1986. Amirante Basin, western Indian Ocean: Possible impact site of the Cretaceous/Tertiary extinction bolide? Geology, 14, 423-426.

3. CARTER, N.L., OFFICER, C.B., CHESNER, C.A. and ROSE, W.I., 1986. Dynamic deformation of volcanic ejecta from the Toba caldera: Possible relevance to Cretaceous/Tertiary boundary phenomena. Geology, 14, 380-383.

4. GILAT, A., HONIGSTEIN, A., 1981. The polyphase history of the Qanaim Valley fault zone. Geol. Surv. Israel, Current Research, 2, 63-68.

6. EMBABI, N.S., El-KAYALI, M.A., 1979. A morpho-tectonic map of the Bahariya Depression. Ann. Geol.Surv. Egypt, 9, 179-183.

7. ROEHLICH, P., 1980. Tectonic development of Al Jabal al Akhdar. In: The Geology of Libya, 3, ed. SALEM, M.J., BUSREWIL, M.T., 923-931, Acad. Press, London, UK.

8. El RAMLY, M.F., ARMANIOUS, L.K. and HUSSEIN, A. A.A., 1979. The two ring complexes of Hadayib and Um Risha south Eastern Desert. Ann. Geol. Surv. Egypt, 9, 61-69.

9. SERENCSITS, C., FAUL, H., FOLAND, K.A., El RAMLY, M.F. and HUSSEIN, A.A.A., 1979. Alkaline ring complexes in Egypt: Their ages and relationship to tectonic development of the Red Sea. Ann. Geol. Surv. Egypt, 9, 102-116.

10. HASHAD, A.H., El REEDY, M.W.M., 1979. Geochronology of the anorogenic alkalic rocks south Eastern Desert, Egypt. Ann. Geol. Surv. Egypt, 9, 81-101.

11. WILSON, W.E., 1976. Famous mineral localities: Saint John`s Island, Egypt. The Min. Rec., 310-314.

Part IV

RIFTING AND DRIFTING

Magma ascends through mid-oceanic ridges along sea-floor spreading centers. In the Red Sea, it is creating a nascent ocean now dilating at a rate of 2 cm a year. This is dividing the African Plate from the Arabian Shield. The former was stationary through most of Phanerozoic time, but is now moving north to ram Europe. The latter is a remnant of the Arabian-Nubian Craton. The effects of these epochal events have been diciphered from magnetic anomaly data in the Red Sea itself.

During the Late Mesozoic, over 70 million years ago, a worldwide sea-level elevation produced a flooding of this kinematic region by an epicontinental sea. There was accompanying tectonism and carbonate deposition. Limestones of the latter provided building blocks for the great pyramids in Egypt.

In the Eocene, luxuriant conditions allowed a wide range of mammals to arise in the Fayum Oasis. Man appeared in the Olduvai Gorge and his spoor radiated south and north, although only the cat-sized, much earlier and possibly ancestral Aegyptopethicus represents him in Egypt before the true man set up stone-tool industries in the Paleolithic.

The accumulation of oil throughout the region was facilitated by the tremendous horizontal scale of the host basins caused by the plate movements and unparalleled elsewhere.

After the Mediterranean dried up toward the end of the Miocene and evaporites were deposited about 5 million years ago, a later inundation occurred. This salinity crisis was accompanied by the climatic change from warm to cold and followed by the onset of aridity.

The first stages of the River Nile coincided with tensional faulting, associated uplift and seismicity both in the sea and in the Gulf of Suez. This Eonile exploited a huge drop in base level resulting from the Mediterranean desiccation to cut an impressive canyon deeper than than the existing Grand Canyon in Arizona and entrenching into bedrock 568 m below modern sea-level near Cairo. The dynamic precursor river conveyed about a quarter of the total mass of the material of the Nile Cone, a proto-delta, in less than three million years. Later, after the Mediterranean refilled, the canyon became transformed into a long estuary.

Chapter IV-7

INUNDATION, UPLIFT AND OIL

"He that toucheth pitch shall be defiled therewith". Ecclesiasticus, 13, 1.

ENERGETIC MARGINS

Why is the Middle East so rich in oil? This question arises because the source rocks are not especially favorable compared with those of other basins. True, the reservoirs are abundant, with high porosities and permeabilities enhanced by fracturing in the Zagros faults. However, they are not in a class apart from the basins of the North Sea or the US Gulf Coast. The effective seals are not unusually thick or abundant. The advantage is that the horizontal scale of the basin is extraordinary so that, while the vertical dimensions are average, the horizontal are greater than elsewhere. The pre-erosional Mesozoic depositional platform was 2,000 to 3,000 km wide and at least twice as long. Differentiation was minimal so that source rocks, reservoirs and seals have a great extent. The vast size of the structures involves frequent closures of 1,000 km² and, due to the tremendous horizontal extension, the resulting structures are very gentle. This has a twofold effect. First, there is a minimal loss of seal efficiency through fracturing and large trap volumes are attached to quite modest vertical closures. Second, because the lithologic facies are so extensive, the horizontal migration of oil is highly efficient. The large structures have a high degree of fill because they could drain very large "kitchens".

Carbonate sedimentation on a stable platform dominated the post- Hercynian sequence of the Middle East and was flanked by the Tethys Ocean. Through time, two main types of depositional system alternated. These were ramp-type, mixed carbonate-clastic and also differentiated carbonate shelf sequences. The former deposited during offlaps when clastics were introduced into the basin. The latter formed during onlaps when carbonate cycles were dominant and separated by lithoclines, diachronous submarine lithified surfaces. There was a marked differentiation with euxinic basins separated by high energy margins from carbonate-evaporite platforms.

The tectonic evolution of the Levant is divisible into several stages. The first, which ended in the Turonian, demonstrated very stable platform conditions with three main features. These were broad regional uplifts, such as horsts, tilted fault blocks and salt domes. All influenced sedimentation. The second, lasting from Turonian to Maastrichtian times, was one of increased orogenic activity. A foredeep formed along the margin of the Tethys followed by ophiolite-radiolarite nappe emplacement.

Figure 7.1: Mobile eastern margin of the Arabian Shield

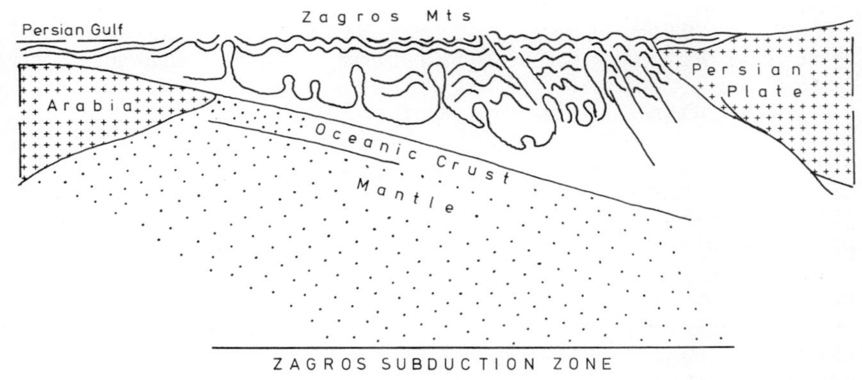

The former was called a geosyncline earlier. However, this term is obsolete and no example of what was meant by it exists on Earth today. Originally it was defined as an elongated depocenter, but became gradually associated and genetically correlated with vast orogenic systems. During the "fixist" era of geology, this might have made sense. However, in this age of mobile plates, the relation between depositional realm and orogeny is much less straightforward. In the Levant and the

Middle East up to the Turonian, sedimentation occurred in an intra- to peri-cratonic setting on a passively northward drifting plate. Here, and for that matter everywhere, the terms "foredeep" and "successor basin" should be used (1).
From the Late Cretaceous to the Miocene, the platform regained its stability, losing it again at the end of the Tertiary. Then, the region was affected by the last Alpine orogenic phase creating the Zagros anticlinal traps. Source rocks formed within the euxinic basins during transgressions and the main reservoirs are composed of marginal mounds, rudist banks, oolite bars and regressive sandstones. The regional seals are supra-tidal evaporites and regressive shales (Figure 7.1).

THE SIRTE BASIN OF LIBYA

From the end of the Cretaceous into the Paleogene, the African Plate drifted northeast, but, during the last 10 Ma, changed to a northwesterly course relative to stable Europe. This induced two tectonic regimes in Tunisia and northern Libya. All this is inferred from stress determinations on homogeneous micritic limestones ranging from Middle Triassic to Quaternary in age. Also, three crustal domains were defined in northern and central Libya, the latter embracing the entire Sirte Basin. Here, the different crustal stress history could be due to a continuous northeastern motion of the African Plate after the ramming of Europe in Late Cretaceous times by the Adriatic promontory. The latter is the continuation of Africa from northwestern Libya and Tunisia to the north.
The Sirte Basin contains fourteen giant oil and gas fields. The Nasser was the first to be discovered and the latest is the Messlah which is estimated to contain more than three billion barrels of oil. It differs from almost all the other Sirte Basin giants in that it is neither a closed structure nor a carbonate "build-up", but rather a clastic stratigraphic termination. The western, up-dip, eroded edge of the Mesozoic non-marine Sarir Sandstone wedges out on to the basement complex. The stratigraphy is divided into two episodes of deposition by an intra-Cretaceous unconformity. Prior to this, continental deposition of 150 m thickness of sediments took place. Thereafter, a range of marginal to deep marine facies reached a thickness of over

2,600 m.

The Early Cretaceous is composed of this oil-producing Sarir Group with sandstones and shales partly equivalent to the Nubian. Following the early Alpine unconformity, the Late Cretaceous is composed of 260 m of Rakb Group mixed sediments. It is succeeded by various facies of the Paleocene, including 16 m of nummulitic limestone at the top of the 570 m thick sequence. The Eocene is mostly made of limestones at the base followed by dolomite, anhydrite and scarcer limestones attaining a thickness of 570 m. The sequence is concluded by Oligocene to Holocene beds attaining a total thickness of 780 m, the lowest part being sands, the middle dolomitic limestone and clay and the uppermost surface sands. In the northern Sirte Basin, subsidence was continuous throughout the Late Cretaceous and Tertiary. It reached a maximum during the Paleocene and Eocene when a major reactivation of faults occurred (2).

During all the Late Cretaceous and Tertiary, shales and carbonates were deposited and abrupt lateral facies changes took place from the platform areas towards the deeper troughs along with steep down-dip thickening. These conditions may have been aided by contemporaneous faulting along structurally weak hinge lines where the dominant structural elements are normal step faults. There is no Late Paleozoic and Early Mesozoic which implies that the area was domed, faulted and eroded during the Late Mesozoic. The Maastrichtian carbonate facies is widely distributed over a great deal of the platform and basinal regions, suggesting that tectonics were rather inactive at this time.

Paleocene crustal extension caused a marked lithological and structural change, shales succeeding limestones in the Marada Trough. Reactivation of earlier faults and an increase of the sediment supply from the south caused these Lower Paleocene shales to blanket the entire area except for the old highs where carbonate deposition went on. An intercalation of shales and carbonates provides a sensitive indicator of depth and sediment changes. This intercalation has been referred to Paleocene cyclic offlapping (3).

Stratigraphic correlation between southern Libya and northern Chad along the margin of the Serir Tibesti facilitated understanding of the Paleocene-Eocene transgression on the African continent. It was discovered that it started in the Late Cretaceous and culminated during the sedimentation of the

Late Paleocene Operculinoides Beds and Early Eocene marls.

Two basement structures have been recognized north of the 26th parallel in Libya. The first comprises the provinces of Central, and northern Libya which are mostly block-faulted and consist of graben and horst structures. Here, Cretaceous and Tertiary tectonic activities produced meridional and SE-NW trending fractures. The basement steepens towards the Mediterranean. The second is composed of west, south and east Libya, a region characterized by a Paleozoic structural configuration made up of a series of basins and uplifts. Most of the oilfields of the Sirte Basin are located on highs of the basement and their presence is related to neighboring graben structures. Oil migrated from the grabens to the horsts along faults (4).

The Amal and Augila oil fields are among the most important. They are located on a high which extends northwest to southeast. Its axis contains Precambrian granites which, during Late Cretaceous time, were exposed at an elevation of 600 m. From the Early Paleozoic until then, the high rose and basement was weathered and eroded, rendering it more favorable for the accumulation of petroleum. Thereafter, the uplift subsided and a marine transgression from the north deposited the clastic beds of the Cretaceous Maragh Formation, an important reservoir rock. The oil itself probably derives from the Rakb Formation which is composed of dark argillaceous limestone containing many fossils such as rudists, crinoids, bryozoans, algae and microfossils.

MOBILE EGYPT

A worldwide elevation of sea-level during the Late Cretaceous and Early Tertiary caused a broad epicontinental sea to transgress much of the northeastern margin of the craton and this culminated in the Ypresian deposition of the Thebes Formation, an extensive mosaic of synchronous carbonate lithofacies. Regional tectonism connected with the Red Sea rifting during the Neogene has produced structural separation and often uplift and removal of much of the Early Eocene in eastern Egypt.

Paleozoic volcanics are found around the Arabian-Nubian Shield. At Gebel Uweinat, a rhyolite flow underlies Early Carboniferous sandstones. Renewed activity in the Late Cretaceous is represented by lavas and tuffs interbedded into basal Nubian sandstones east of Kom Ombo.

There was further volcanicity in the Oligocene connected with the instability in the Red Sea area premonitory to its Miocene dilation. In fact, along this coast, a dolerite flow occurred at Quseir. Also, in the Cairo-Suez area, there is a basalt belt extending as far as the Fayum. Many dike swarms are found in Sinai and usually they relate to fault lines. It is almost sure that the active magma of the Tertiary volcanoes of Lower Egypt was basalt, but, despite its homogeneity, the eruptions range from effusive to highly explosive and even cryptovolcanic. In addition, the lava flow at Gebel Qatrani north of Fayum is partly due to fissure eruption.

A remnant of the sediments deposited along the eastern margin of an epicontinental sea which once covered most of Egypt is preserved in the Duwi Trough on the existing Red Sea coast. The Paleogene in the Nile Valley has long been regarded as the type section for the entire Eastern Mediterranean. Its Paleocene Esna Shales have been used for making pottery since antiquity.

Figure 7.2: A butte made of Dakhla Shale west of Abu Monquar, Western Desert.

Near the Kharga Oasis on the Abu Tartur Plateau capped by Thanetian Kurkur Limestone, there are two groups of Landsat-detected fractures. The first affects the semi-circular slopes of the plateau cutting "Nubia", Duwi and Dakhla Formations, while the second exerts its influence on the overlying rock (Figure 7.2). Both run longitudinally with occasional latitudinal deviations controlling drainage along the slopes. The radial distribution of both fracture and drainage lines favor the concept that the pattern results from vertical movements which promoted updoming in the Late Cretaceous and Early Eocene.

The Thebes Formation showed a three stage depositional history. Initially, there was a basin-wide pelagic carbonate deposition with thin, shallow water facies near the margins. Then, a gradual shallowing occurred with the establishment of widespread benthonic communities during intermittent pelagic and simultaneous tectonic uplift of fault blocks in the east accompanied by the production of small carbonate platforms separated by deeper basins. Finally, an abrupt lowering of sea level was associated with the development of shallow water organic build-ups within the basin and exposure together with erosion of uplifted fault blocks to the east.

Figure 7.3: The Sphinx at Gizeh carved from Eocene nummulitic limestone. In the background is the Great Pyramid.

The Eocene is composed of limestones, clays and sands, the limestones comprising the Mokattam Formation. The Priabonian comprises two major sedimentary sequences with unlike faunas and different paleogeographic histories. The lower one is nummulitic with Nummulites gizehensis, etc. It corresponds to a regional regressive episode. The upper one is mainly characterized by the frequent recurrence of the index pelecypod Carolia placonoides.

The Pyramids at Gizeh were constructed using blocks quarried in the Mokattam Escarpment. Herodotus described them and recorded the history of their construction as related to him by the priests. He was the first to mention nummulites which he believed to be preserved lentils of slave meals (Figure 7.3).

In the Shabrawet Hills south of Ismailia and adjacent to the Great Bitter Lake, the sequence from the Early Cretaceous to the Quaternary is upwards of 400 m thick and includes clastics and carbonate rocks with several unconformities. The Late Cretaceous rests unconformably on preceding Albian and is 270 m thick. There is an unconformity in the Campanian and another at the base of the overlying Middle Eocene. The marine Eocene is 240 m thick and split by two unconformities during and at the end of the Lutetian. Also, it terminates with an unconformity separating it from the continental Oligocene. The sequence ends with marine Miocene limestones having unconformities below and above, gravels and eolian sands.

Along the continental margin to the north of the neighboring Sinai Peninsula, there are many diapirs usually aligned along westerly faults. These come from a Late Tertiary evaporitic layer. A 90 km long salt ridge produces a submarine uplift in the lower continental rise. Two main fault trends occur. They are parallel to the continental slopes of northern Sinai and southern Israel, respectively. The structural pattern of the Sinai margin is probably controlled by two major tectonic elements. These are rejuvenated basement faults of the continent-ocean transition zone and salt diapirism due to the loading of Messinian evaporites with Nile-derived Late Tertiary clastics (5).

OIL IN THE DESERT

Oil was first discovered in the Western Desert in 1966 at El Alamein where, by an irony of Fate, Field

Marshal Rommel lost the decisive battle to General Montgomery in 1942 for lack of it. Four commercial oil fields have been established there. Some yield both oil and gas, nearly all the horizons belonging to the Cretaceous. However, there is an oil-bearing sand in Um Baraka which could be older and oil and gas shows have been found in many wells in formations ranging from Jurassic to Eocene. The Mesozoic and Early Tertiary rocks are good potential sources for hydrocarbons because they possess organic-rich sediments, abundant reservoirs and numerous caprocks.

Lower Cretaceous Aptian sediments are mostly clastics succeeded by carbonates. Structurally, they were affected by the Atlas tectonic trend of the early Alpine Orogeny. The later pulses of this coincide with the Laramide Revolution of the Late Cretaceous to Early Tertiary in the Syrian Arc. A large-scale Cretaceous-Tertiary hiatus suggests that a Syrian Arc movement reinforced northern ridges such as the Qattara-El Alamein high in which the oilfield occurs, uplifting them. However, in the case of the Ras Qattara-Tiba high, subsidence began at the end of the Cretaceous and appears to be unconnected to the Syrian Arc movements. Faulting is evident everywhere in the area, this producing a horst-graben fault pattern along the Qattara-El Alamein high. The Aptian Sea seems to have encroached from the north over successively older stratigraphic levels down to the Paleozoic. The residual highlands provided clastics and were submerged in the Late Cretaceous. Shallow water shelf deposition occurred on seamounts at Abu Roash and Wadi Natrun. The west Qattara area remained structurally low and a semi-closed Aptian basin developed. Further north, sedimentation occurred in a shallow fluctuating sea, but, at Matruh, there was an abrupt increase in the rate of sedimentation accumulating the Matruh Shales of 3,000 m thickness. The Matruh Basin has an excellent oil potential. At the end of the Aptian, a transgression covered most of the northern Western Desert. As the sea partially regressed, a transitional environment was resumed and dolomitization was activated. The resulting surface is seismically indicative as a marker. The information regarding petroleum in the region has underlined the significance of Syrian Arc-type structures as traps for oil migrating from primary reservoirs.

In the Abu Gharadig and Fayum Basins, there were probably extensive marine transgression and regression cycles and deposition combined with a minimum of three tectonic cycles. A highly deformed thick sedimentary sequence resulted which, from the Jurassic to the Lower Tertiary had its organic-rich material subjected to sufficient heat and pressure to generate hydrocarbons. These high quality oil and gas reserves make these basins potentially the greatest hydrocarbon reserves in the Western Desert of Egypt. They are separated structurally by a major ridge and both are bounded to the north by another structural high associated with Cretaceous and Tertiary faulting. The southern boundary comprises an extensive platform with a thin sedimentary cover. Deep-lying groundwater has flushed the southern parts of these basins, consequently there is little chance of hydrocarbon preservation there. Elsewhere, however, there is a good hydrocarbon potential in cyclic sequences of sandstone, shale and carbonate with moderate porosities and permeabilities (6).

Regarding the oil shale potential in Egypt, Late Cretaceous black shale outcrop samples and cores, all weathered, except those for mines, produced oil yields reaching 86 l/t with an average gas content of 3% by weight. An estimate of all potential organic-rich strata amounts to 4.5 billion barrels of oil in situ in the area Safaga-Quseir and, at Abu Tartur, 1.2 billion barrels of oil can be reckoned on. In the Eastern Desert, block faulting and steep dip of the shales inhibit commercial oil exploitation. Neverthelesss, they may be also of use for the production of cement or for direct combustion in an electric power plant (7).

OIL IN THE RED SEA

In the Gulf of Suez during the Miocene, thick evaporites were deposited, reflecting closed basin conditions and sealing in the oil. Afterwards, its receding margins sank and a rampart of reefal carbonate accumulated. These evaporites may have been precipitated in the Early Miocene. If this is confirmed, then the Mediterranean Messinian desiccation must have begun earlier in its peripheral gulfs and embayments.

This Gulf is a good example of intracratonic rift basin evolution. It is by far the richest oil province in Egypt and petroleum accumulations are most frequent in the Miocene, although not confined to this Epoch. In the last decade or so, scores of wells were drilled every year, hence many large oil fields have been discovered. Up to 1984, almost 3.5 billion barrels of petroleum were produced, nearly three quarters from the Miocene sediments of which the Kareem Sands are the main producer. Of the rest, most comes from the "Nubian Sands". Late Cretaceous and Paleogene sediments were deposited in a marine platform environment in the region.

In the Early Eocene, regional uplift caused a retreat of the sea which persisted through this Epoch. Later, mostly during the Oligocene, tensional faulting promoted the subsidence of the Gulf of Suez graben with its associated horsts. Subsequently, the latter were severely eroded. The Miocene sea invaded from the north, submerging older structures unconformably and filling the graben. Its deposits accreted in residual basins delineated by upthrown blocks within the Gulf. More than 4 km of these overlie pre-rift basement and sedimentary rocks and they comprise initial clastics and succeeding evaporites. Despite the relatively short period of deposition, the total Miocene thickness of sediments is twice that of pre-Miocene, post-basement rocks.

During the Early Paleogene, a complicated pattern of mainly carbonate and lesser associated terrigenous facies was deposited in narrow, elongate tectonic basins or in Tethyan branches. The basins were separated by structural ridges connected with the emergence of the Syrian Arc. Through the Eocene, infilling of pre-existing continental depressions provided a depositional base for seward progradation of continental sediments. Through the Oligocene and Miocene, progradation was from southwest to northeast over an open shelf and it resulted in deposition of a thick terrigenous section in the Western Desert. Longshore currents helped redistribute sediments towards the present Nile Delta. The result was an eastward shift of terrigenous centers of deposition from Oligocene through Miocene time. As this took place, carbonate environments reappeared in the west, onlapping abundant parts of the fluvial delta systems. By the Early and Middle Miocene, carbonate environments became extensive and continued to migrate eastward. It is uncertain whether the Late Miocene is represented in Egypt or not, but clastics near Lake Manzala in the eastern Nile Delta may be of this age. A branch of the Mediterranean initiated development of the Gulf of Suez during the Oligocene and Early Miocene. It was the site of deposition of arkosic sands, fanglomerates and some evaporites. However, fine-grained terrigenous sediments were

transported by longshore currents to the Gulf of Suez from early Nile deltas near the present one. Large volumes of evaporites, contemporaneous with the Nile Delta accumulated in the Gulf of Suez basin during the Middle and Late Miocene as a result of a clastic fan at the north end of the Gulf. The Miocene Globigerina Marls comprise a significant part of the terrigenous sediments in the Gulf of Suez, are thought to have been derived from longshore currents from the Nile and are considered to be a source rock for hydrocarbons (8).
Structurally, the Gulf of Suez is divisible into three sub-basins separated by a couple of discontinuous ridges running NW-SE parallel to the present Gulf. The eastern one, the Abu Durba ridge, was mostly subaerial, but the western Zeit-Gharib ridge has been submerged at least since Middle Miocene times. As the sea invaded from the north, the sediments thicken in that direction. Along the eastern side of the Gulf of Suez, there are lateral trends of chemical variation in fluid components. These variations are believed to be due to chemical interactions between oil, water and country rock.

THE OIL LEVIATHAN

After a disconformity which ended the Cretaceous, in Saudi Arabia a widespread Paleocene transgression resulted in the deposition of the Umm er Radhuma Formation. This is a thick succession of neritic limestones and more basinal marls, the former reaching 240 m at the type section. Similar rocks also covered most of Jordan. The Ypresian age ushered in persistent, widespread evaporite deposition to considerable thicknesses in the western Persian Gulf, southern Iraq, Kuwait and Qatar.
Geologically, the surface of Qatar comprises Tertiary limestones and dolomites with interbedded clays and marls covering 80 % of the Peninsula. Quaternary surficial deposits of conglomeratic limestone, sands and sabkhas cover the rest. The predominant rock unit is the limestone of the Middle Eocene Dammam Formation, the Early Eocene Rus Formation being restricted to the western coastal zone south of Dukhan. Middle and Early Miocene marls and sandy limestones constitute the Dam Formation and are capped by the Mio-Pliocene Hofuf Formation with calcareously cemented conglomerates. The largest single exposure of this is superimposed on the

Ghawar Field, perfectly mirroring the outline of that giant oil-filled anticline (9).
The Early Eocene evaporites on the eastern flank of the Arabian-Nubian Shield constitute the Rus Formation which is 55 m thick at the type locality. Following it is the Dammam Formation, 30 m thick, consisting of limestone, dolomite marl and shale. The beds are petroliferous and reflect a marine invasion with both nummulitic and globigerinal carbonate deposition. Supporting micropaleontological data from the Middle and Upper Shumaysi Formation from Usfan near Jeddah are also compatible with an Early Eocene regression followed by a moderate transgression in the Red Sea.

Figure 7.4: Seismic belt encircling the Arabian Shield (coarse dotting represents greater activity). Polar wander path for Arabia and Africa are shown for post-Cretaceous time.

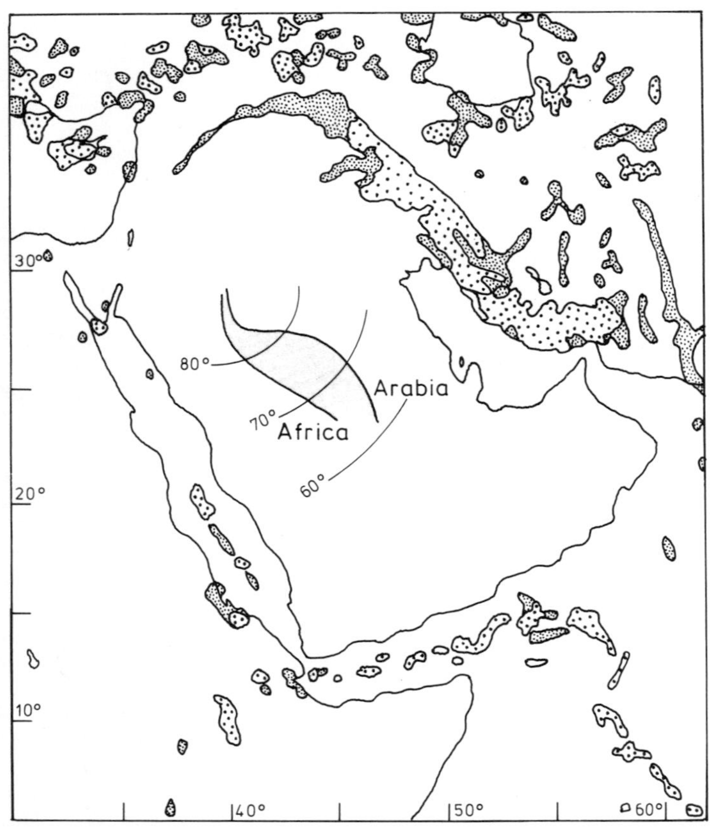

Inundation, uplift and oil

There was a widespread emergence of the Arabian platform in the Middle Eocene which reduced the Tethys Ocean to a relic sea. This has endured until now, except for sporadic flooding of the present coastal area in the Middle Miocene. The initial formation of the succeeding Mio-Pliocene is the basal Hadrukh, calcareous silty sandstones and limestones which are 85 m thick at the type section and lie unconformably on the Dammam. Overlying this is the Dam Formation, 90 m thick, composed of marls and shales with beds of limestone and sandstone together with coquina. Above this is the Hofufate Formation, 95 m thick at the type section, comprising sandy marls and limestones with subordinate calcareous sandstone and local gravel beds toward the base. Finally, the Kharj Formation, 30 m thick at the type section, includes marine and lacustrine limestones, gypsum and gravel. Huge lava fields in the shield and northwestern Arabia attest to extensive volcanism at this time. Surficial deposits, gravels, sands and silts, complete the geological succession in Saudi Arabia.

Paleomagnetic poles were determined for three sequences of Tertiary basalts and also for the Usfan Formation (latest Cretaceous to Paleocene) near Jeddah in western Saudi Arabia. The lavas, which range in K-Ar ages from 29 to 5 Ma, pertain to the opening of the Red Sea. The paleomagnetic pole for the Early Usfan Formation, aged 70 Ma, was on the sixties latitude north and two hundred and thirties longitude east. Thereafter, it shifted consistently northward until reaching its present position. The poles from rocks older than 90 Ma demonstrate displacement from the apparent polar wandering path of the African Plate, reflecting the opening of the Red Sea Rift System from 29 to 24 Ma ago with a counterclockwise motion about a rotation pole (Figure 7.4). The northward movement of the Arabian Plate from the Maastrichtian until the Early Miocene averaged 5 cm annually, i.e. an overall movement of 2,400 km. By then, Arabia was almost in its present latitudes, impelling Turkey southwestward in the direction of Africa. From the relationship between extensional, compressional and strike-slip deformation, it was concluded that the Turkish-African convergence results in complex crustal welding in a broad and diffuse zone rather than in active consumption of African lithosphere under the Cypriot-Tauride Arc (10).

Continental margins may be Atlantic or Pacific in type, according to their origins. The former undergo steady sinking and accumulate enormous thicknesses of sediment. They may be characterized as passive. On the other hand, the latter are usually rising and associated with volcanism, folding, faulting and other orogenic processes. They comprise active regions.

The Red Sea belongs to the Atlantic type and started from the rupture of an ancient continent along a line of stress. There was later subsidence and loading of sediment. Here, the upwelling sima actuated the rending of the Arabian-Nubian Shield, separating the Arabian Peninsula from Africa. The edges of the opposing continental margins tend to break off and sink as long narrow blocks. Reefs can grow on these and build up carbonate shelves which further depress the crust through their weight.

The western part of the Arabian Plate as far as 300 km toward its interior from the Red Sea Rift System is an active tectonic region with high risk of a damaging earthquake on the western fringe of Saudi Arabia where intensive urban and industrial development is progressing. For instance, two significant earthquakes occurred in the southwestern part of the kingdom in 1941, one in January near Jizan, where salt diapirism is apparent, with a magnitude of 6.25 on the Richter scale and another in February with a magnitude of 5.5. A further earthquake of magnitude 5.5 took place in October 1965 near the Saudi-Yemeni border. A sequence of earthquakes was recorded in March 1967 along the Red Sea Rift System only 150 km southwest of Jeddah. Arabic historic records include the famous work on earthquakes, "Kashf as-Salsalahan wasf Azzalzalah", by As-Soyuti written about 1500 (11).

By 1982, Saudi Arabia alone had produced over 44 billion bbl of oil and had remaining reserves estimated as in excess of 177 billion bbl. These vast quantities imply a very efficient setting for petroleum generation as well as effective migration, collection and preservation. It is suggested that, in the future, perhaps 50 billion bbl more of producible oil may be discovered and total production could rise to more than 271 billion bbl, corresponding to 800 billion bbl in the reservoir. Even assuming a 25% loss of oil through faults or because it was not originally trapped, the total petroleum generation in the area could exceed a trillion bbl (12).

The age and location of the source rocks of the oil is still speculative. Iranian geologists proposed an Eocene source. Others believed that the enclosing

Jurassic carbonates were important when the Zagros Mountains underwent overthrusting, acting like a rolling pin to force oil into shelf limestones from the east. The Zagros Mountains underwent active tectonogenesis since the Paleozoic and their structural characteristics are explicable mainly by vertical movements with the Hormuz Salt acting as a means of halokinetic uplift and folding. It is improbable that Tertiary orogenic thrusting and shearing acted as the sole or primary originator of the present-day structures (13).

Since 1900, over sixty structurally controlled oil and gas fields have been discovered and most are huge, multi-reservoir accumulations which produce from fractured carbonates ranging from the Permo-Triassic to the Oligo-Miocene.The main source of the oil is the Albian Kazhdumi Formation. It is suggested that this source rock was not buried deeply enough to generate hydrocarbons until the Eocene, also that no important oil expulsion occurred before the Miocene.

However, it is tenable that the Middle Cretaceous off-shore oilfields of the Persian Gulf derived their petroleum by lateral migration from an Albian source rock. If this is true for the Burgan and other oil fields in Kuwait also, problems regarding the concept of vertical migration from an older Berriasian-Valanginian place of origin can be avoided. However, at Burgan there is a central bitumen seep and subsurface information clearly shows that the oil does in fact derive by vertical migration from the underlying Eocene formations which are saturated with tar and heavy oil. The mechanism was very likely through faults, some oil reaching the surface and weathering to bitumen. At Magwa, there are gas seeps overlying Eocene gas and tar accumulations and probably escaping through faults.

Prior to the deposition of the Middle Cretaceous Mauddud Limestone in the Greater Burgan Oil Field, this area of Kuwait was part of an extensive sea. In this, regressive sedimentation took place and is recorded by the underlying Burgan Sandstone. Subsequently, the sea inundated the shelf, commencing the deposition of the Mauddud Formation in the northern Persian Gulf and the Bangeston Limestone of the Zagros Mountains. Most of the Mauddud Formation in Kuwait originated as very shallow lagoons, marine and tidal flats interrupted by occasional influxes of clastics. After a later hiatus, the regressive Wara Sandstone was laid down.

However, in the event that vertical migration is to be discounted for the Middle Cretaceous fields, the insistence on an older and deeper Callovian-Oxfordian provenance for the enormous amount of oil in Saudi Arabian reservoirs, the greatest petroleum accumulation so far discovered, is questionable.

In 1982, valuable geochemical data for source rock maturity became available. From them, it is concluded that the only rocks rich enough to have originated the gigantic reserves of Saudi Arabia were indeed the Jurassic, actually the Callovian to Oxfordian carbonates. These have a total organic content averaging 3 to 5%. The estimated volume of the mature source rocks and the actual output of oil do not agree. This may be due either to error in the organic geochemical parameters or contributions from other source rocks which may lie outside the area.

In eastern Saudi Arabia, the Harmaliyah Field comprises a slightly younger Kimeridgian reservoir possessing a relatively high primary porosity. It seems that the host calcarenites retained this characteristic during Kimeridgian to Cenomanian time, i.e. before any possible hydrocarbon entrapment. Therefore, other mechanisms such as shallow burial must have acted to preserve the porosity.

The Late Cretaceous Omani foredeep extends along the Trucial Coast and formed during the Campanian and Senonian. The associated deposits enlarge abruptly from less than 300 m on the western shelf to more than 1 km in the trough. In the latter, shales and argillaceous limestones of the Campanian are said to lie unconformably on Turonian carbonates. Downwarping continued after the close of the Period with accompanying deposition of Tertiary sediments. In the Paleogene, the deposits were mainly shale, marl and limestone. The development of a closed basin environment during the Neogene caused a substantial accumulation of anhydrite and halite in association with marl and shale. In the last stages of infilling of the trough, the aridity extended beyond the confines of the foredeep and onlapped parts of the shelf to the west.

The Fahud Field lies on the western hingeline between the shelf and the Omani part of the foredeep. Further north and northeast, other hydrocarbon accumulations occupy similar geological settings such as the Musandam Peninsula. The deep trough sediments might have generated hydrocarbons which later migrated to accumulate in the shelf edge reservoirs. However, in the case of the Fahud Field, the relevant sediments were too immature to do so.

Instead, the hydrocarbons may have derived from Paleozoic or older Mesozoic sources, suggesting emplacement by vertical migration during the Late Cretaceous and Eocene.
Additionally, the shallow shelf carbonates of the Middle Cretaceous Mishrif and Mauddud Formations in the Fahud Field and throughout northwestern Oman contain reservoir facies. In the Mishrif, porosity developed in sand-aprons of lithoclast and skeletal grainstones surrounding fault-block islands. It occurs less commonly in the Mauddud as biostromes of rudist packstones.
On the western margin of the Omani foredeep, the lowest formation of the Middle Cretaceous is a shale which seals in oil and gas accumulations of one reservoir. However, there is another separate, quite distinct, second reservoir in which the crude oils are dissimilar. The presence of vertically separated and sealed major gas and oil reservoirs in other fields on the same hingeline makes vertical migration improbable.
Alternatively, petroleum generation and migration may be an integral part of diagenetic changes affecting a limestone section undergoing subsidence. Stylolite formation in the carbonate is critical in hydrocarbon expulsion. The onset of stylolitization at Fahud roughly coincided with the time of emplacement of oil and gas in the porous rocks of the Middle Cretaceous Wasia Group suggested above.

FRAGMENTED LEVANT

Since the Cambrian, the Near East underwent epeirogenic oscillations, these being of greater effect in the unstable shelf than in the stable one and causing six large-scale transgressions. There were vertical differential movements occurring contemporaneously which created the structural division into troughs and swells in the unstable shelf. These relate to active deep-seated structural zones, the strongest subsidences occurring at areas were they intersect. Within the unstable shelf, increasing consolidation is perceptible since the Silurian. This affects much of the formerly tectonically unstable area and the most important of the earlier troughs were transformed into uplifting areas during the Late Cretaceous and the Early Tertiary.
The Early Cretaceous Gevaram Shales occur in the subsurface of the western coastal plain and offshore

of Israel. From vitrinite reflectance measurements and the organochemical characteristics of cores, it appears that the degree of maturation increases westward where thermal generation of oil responded to the increase of burial depth. The Turonian-Senonian may be the earliest time at which the relevant physical conditions existed. In the east, the maturation did not progress much after this time. However, there was a further increase in the degree of maturation to the west during the Neogene because of the thicker accumulation of sediments. In the Late Tertiary of the coastal plain of Israel, a highly unsaturated, sulfur-rich, naphthenic condensate occurs in association with dry, biogenic gas. It is believed that this condensate generated in thermally immature rocks from resinous, continentally derived organic matter (14).

From the Late Eocene to Early Miocene, Israel was above the sea. The uplift was intense and rapid. This event was succeeded by a period of erosion and gradual, regional subsidence. It led to the flooding of the inland basins from west to east by the earliest Middle Oligocene. Therefore, a new depositional marine sequence began and was bounded basally by an unconformity. Erosion continued on lands which were still emergent.

On a larger scale, major worldwide falls in sea-level can be inferred from significant regional unconformities developed by subaerial erosion. It is thought that the greatest drop in sea level since the Trias took place at the beginning of the late Middle Oligocene. Hence, the Israeli uplift alluded to above may not necessarily be local, but a part of a global phenomenon (15).

Calcareous bituminous rocks of the Campanian-Maestrichtian Ghareb Formation are exposed on the margin of the Dead Sea graben, on downwarped blocks, and probably extend under the thick fill of the graben. The agreement of the geochemical indicators implies that these bituminous rocks constitute a potential source of the asphalts and oils in the area.

In the Dead Sea Basin, asphalt occurs in three varieties, as blocks up to 100 tons in weight and practically pure asphalt, veins or seepages together with fissure fillings in Early Cretaceous to Holocene rocks and ozocerite veins on the eastern shore. The oil and asphalts belong to a single geochemical province and their properties suggest a source material of bacterially reworked algae and a source rock poor in clay minerals, probably a

calcareous rock (16).
The surface manifestations interrelate with tectonic activity and in the lake itself, diapirism is linked to the asphalt occurrences. Hydrogen sulfide gas and semi-liquid asphalt are sometimes released by a combination of this and tectonism. During periods of minimal crustal mobility, as in the last and present centuries, the rate of seepage of asphalt to the Dead Sea bottom sediments is considerably slower and asphalt solidified on the bottom of the lake. Its release to the surface became more spasmodic, possibly, in part, resulting from earthquakes. Interestingly,the first recorded war for control of a hydrocarbon deposit was in this area and took place in 312 B.C. between the Seleucid Syrians and the Nabatean Arabs.
In Jordan, basin and uplift alternation persisted through the Paleocene and most of the Eocene following its initiation around the Wadi Araba stress region in the Oligocene. There, coarse clastics derived by erosion of peripheral areas were deposited. In the center of the graben there are local rock salt accumulations which formed from this period until the Pliocene, but elsewhere carbonaceous clays and fine clastics formed. During the Neogene, the paleogeographic situation depended upon the relationship between subsidence due to drifting and the augmentation of sediments. The rift either had fresh water lakes draining into the Tethys, formed an inland basin or underwent a marine transgression from the north. At some time, the Wadi Araba might have comprised part of a link between the Mediterranean and Red Seas which could have also involved the Gulf of Aqaba and the Dead Sea.

REFERENCES

1. IBRAHIM, M.W., KHAN, M.S. and KHATIB, H., 1981. Structural evolution of Harmaliyah oil field, eastern Saudi Arabia. AAPG, Bull., 65, 2403-2416.

2. SCHÄFER, K., KRAFT. K.-H., HÄUSLER, H. and ERDMANN, J., 1980. In situ stresses and palaeostresses in Libya. In: The Geology of Libya, 3, ed. SALEM, M.J., BUSREWIL, M.T., 907-922, Acad. Press, London, UK.

3. GUMATI, Y.D., KANES, W.H.,1985. Early Tertiary subsidence and sedimentary facies - northern Sirte Basin, Libya. AAPG, Bull., 69, 39-52.

4. MIKBEL, S.R., 1981. Major basement structures and their relation to some oil fields in Libya (North of Lat. 26oN). Z. deutsch. geol. Ges., 132, 547-554.

5. BEN-AVRAHAM, Z., MART, Y., 1981. Late Tertiary structure and stratigraphy of North Sinai continental margin. AAPG, Bull., 65, 1135-1145.

6. AWAD, G.M., 1984. Habitat of oil in Abu Gharadig and Faiyum Basins, Western Desert, Egypt. AAPG, Bull., 68, 564-573.

7. TRÖGER, U., 1984. The oil shale potential of Egypt. Berliner geowiss. Abh., A, 50, 375-380.

9. YEHIA, M.A.A., 1983. Atlas of Qatar from Landsat images, 166 pp., Ad Dawhah, Qatar.

10. HARSCH, W.,KUEPFER,T., RAST, B. and SAEGESSER, R., 1981. Seismotectonic considerations on the nature of the Turkish-African plate boundary. Geol. Rdsch., 70, 368-384.

11. YOUSIF, I.A., BECKMANN, G.E.J., 1981. A paleo-magnetic study of some Tertiary and Cretaceous rocks in western Saudi Arabia: Evidence for the movement of the Arabian Plate. Bull. Fac. Earth Sci. Univ. Jeddah, 4, 89-106.

12. AYERS,M.G, BILAL, M., JONES, R.W., SLENTZ, L.W., TARTIR, M. and WILSON, A.O., 1982. Hydrocarbon habitat in main producing areas, Saudi Arabia. AAPG, Bull., 66, 1-9.

13. ALA, M.A., 1982. Chronology of trap formation and migration of hydrocarbons in Zagros sector of southwest Iran. AAPG, Bull., 66, 1535-1541.

14. NISSENBAUM, A., GOLDBERG, M. and AIZENSHTAT, Z., 1985. Immature condensate from southeastern Mediterranean Coastal Plain, Israel. AAPG, Bull., 69, 946-949.

15. MARTINOTTI, G.M., 1981. An Oligocene unconformity and its interregional interest. Current Research, 2, Geol. Surv. Israel, 30-35.

16. SPIRO, B., WELTE, D.H., RULLKÖTTER, J. and SCHÄFER, R.G.. 1983. Asphalts, oils, and bituminous rocks from the Dead Sea area - a geochemical correlation study. AAPG, Bull., 67, 1163-1175.

Chapter IV-8

MAMMALS IN A GARDEN OF EDEN

"We do not know what thoughts stirred in the mind of the last of the mastodons, but we can take it that they were nothing very remarkable".
Anatole France, Under the Rose, 1925.

THE SPOOR OF MAN

The origin of humanity was explained by the ancient Egyptians in the form of curiously fascinating legend embodied in their mythology. It seems that Atum, the God of Heliopolis, emerged from Nun, either by an effort of will or as its child, and fathered a son, Shu, and a daughter, Tefnut, who got lost in Nun and were found by his separable eye. After they were reunited, Atum wept for joy and from his tears grew men. Thereafter, Atum left the waters of Nun and created the Earth, Geb. This was accomplished through the union of his children, who also created the sky, Nut.

More than a century ago, Charles Darwin speculated, without the benefit of fossils, that human evolution was initiated in Africa. The earliest age of Man in Egypt is intimately linked with the extraordinary archeological and anthropological discoveries made by the Leakeys in and after 1959 at Olduvai Gorge in Tanzania. This is a 40 km long river-carved canyon which drops abruptly in the Serengeti Plain and provides well-exposed strata reaching back almost 2 Ma. It was an austere Garden of Eden, hardly as serendipitous as that described by the Pharaoh Akhenaten more than three millenia ago in his great hymn to the Aten. Here, he depicted "the entire Earth performing its labors. All cattle are at peace in their pastures. The trees and herbage grow green. The birds fly from their nests, their wings beating in praise of Thy spirit. All animals gambol on their feet, all the winged creation live when Thou hast risen for them".

The Leakeys started working in the lowermost bed, nearly 100 m below the rim of the canyon. An almost 2 Ma old large hominid fossil was found and called Zinjanthropus boisei. This means East African Man. There were also thousands of animal bones and crude stone tools preserved by solidified volcanic ash. "Zinj", now redesignated Australopithecus boisei, had such massive teeth that he was termed Nutcracker Man. He constitutes a dead end branch of the hominids as does Paranthropus robustus too, a possible relative from Sterkfontein, South Africa. Another contemporary was Homo habilis, the Handy Man from Lake Turkana in northern Kenya. His fragmental remains occurred near those of "Zinj" and they were either neighbors or competitors. By K-Ar dating, it was determined that this, the DK site, was 1.75 Ma old and constitutes the oldest known "human" living floor on Earth. The stone artifacts, including piercing, cutting and scraping tools, comprise the Olduwan Industry. A circle of lava blocks some 4 m across and roughly a third of a meter high is believed to be the oldest building ever found. This industry changed over the next million years to become the Advanced Olduwan between 1 Ma and 750,000 years ago. Associated creatures built crude shelters on the shores of the ancient lake at Olduvai, stood from 1.2 to 1.5 m tall and had the basic attributes of humanity. That is to say, they had large brains and bipedal locomotion. Their fossil remains have been unearthed also in Uganda and Ethiopia. There are at least two or perhaps three types which existed simultaneously in the Late Pliocene and Early Pleistocene from 5 to 1 Ma ago. Vegetarian forms resembling Zinjanthropus are termed Australopithecus robustus. Large in size, they died out early. A smaller form, Australopithecus africanus, a hunter as well as browser, may have been ancestral to Man. Homo habilis is another candidate and had a bigger brain. His globe-trotting descendant was Homo erectus, Upright Man. An ancient form of this, found together with the elephant Loxodon atlanticus, could have been the telanthrope resembling the Tchadanthropus from Chad, this old names now having been discarded. Nevertheless, Chadian Man remains the earliest Saharan hominid. Upright Man could be considered intermediate between australopithecids and pithecanthropids of the Early to Middle Pleistocene. In the same category are included Java Man and Peking Man. Probably due to population pressure, they entered Asia and their bones have been found in India. Perhaps 300,000 years ago, they were replaced by Homo sapiens. Neither australopithecines nor Olduwan implements have been found in the Nile Valley of Egypt and the Sudan. This is also true for Libya and the Arabian Peninsula. Homo erectus and Homo habilis are missing from this vast area as well. However, Neanderthal Man has been recorded at Haua Fteah near Benghazi in Libya and furthermore in the Sudan at a pre-pottery Mesolithic site near Wadi Halfa. Archaic Homo sapiens, who lived between 300,000 and 35,000 years ago, was discovered at Singa in the same country.

Among the close relatives of Man are said to be the pongids. One of the earliest of them is the largest Oligocene primate, Aegyptopithecus zeuxis. Represented by a 33 Ma old partial skull with dentition, isolated teeth and jaws, it closely resembles the Miocene pongid Proconsul, an East African subgenus of Dryopithecus. In size, it is similar to both Propliopithecus and Limnopithecus. In fact, a jaw fragment of Aegyptopithecus has been recorded from the Miocene in Rusinga Island, Lake Victoria, Kenya. All of these dawn apes were only about the same weight as a modern human infant. They had monkey-like limbs and ape-like teeth used to eat fruit and must have been rather cat-like in appearance. Aegyptopithecus is the oldest direct ancestor of Man. Notwithstanding, there is no connecting link between this dawn ape and the almost 30 million years younger Australopithecus. Among proposed intermediates are two from Kenya, Proconsul and Kenyapithecus, together with another pair from there, Rama-pithecus and Sivapithecus. As their names imply, they occurred also in India and China. Finally, there is one from Europe which is Rudapithecus. These ape-like creatures lived at various times between 20 and 8 Ma ago (1).

As regards bones of australopithecines, an adult female, 1.1 m tall and probably weighing 30 kg, was reconstructed from hundreds of pieces of bone. This followed the observation of an arm bone found in Pliocene sediment and volcanic ash in the Afar Badlands of Hadar, Ethiopia. The reconstructed female rapidly became known as Lucy after a Beatles song, but her scientific name is Australopithecus afarensis. Actually, she is related to the even better known Taung child. This infant was born between 1 and 2 million years ago on the high veld in the Transvaal and probably died in a cave where her bones were petrified and thus preserved.

Aegyptopithecus may be in or near to the ancestry of all Catarrhini or alternatively, of all the later Hominoidea, including the gibbons. This is not justified in the case of the Fayum primates. Since they do not resemble the Old World monkeys and must have been separated from both the New World monkeys and the hominoids. Such a splitting of the New and Old World primates, Catarrhines from Platyrrhines, is impossible later than the Early Eocene 50 Ma ago. This is because migration of arboreal primates between the New and Old Worlds could not have taken place later.
It is improbable that Aegyptopithecus was kin of both the Greater and the Lesser Apes. To justify such an assumption, it would be neccessary to show that the dental anatomy of Aegyptopithecus resembles that of Limnopithecus and Proconsul from the much younger East African deposits. This is not the case, its dentition being much closer to primitive Dryopithecus species. However, there are still differences from these in that the size is absolutely smaller and the teeth more broadly transverse. There is no evidence that Aegyptopithecus survived into the East African Miocene. At this time, the diversification of both the Great Apes and the Lesser Apes was well in progress.

PALEOGENE MAMMALS

There are practically no mammals in either the Paleocene or Eocene of Africa, but they can be found in the Early Oligocene of the Egyptian Fayum. Unfortunately, this is at the northeasternmost end of this continent and there is evidence that the fauna in question immigrated from Eurasia. Among its members are many anthracotheres, hyaenodont creodonts and anomaluroid rodents, all which belong to Eurasian groups and therefore may have invaded the region. There were some autochthonous forms of which the most famous are Arsinoitherium and proboscideans which include the archaic Moeritherium and two mastodon genera. However, there are also indigenous hyracoids.
The evolution of the mastodonts, elephants and their relatives is perhaps one of the most exciting developments in mammalian history. They belong to the Proboscidea which, as implied above, is African in origin, but successfully spread to other continents. Consquently, they were widely distributed

in Eurasia and the Americas by the Middle Cenozoic. As early as 1899, it was suggested that Africa was their homeland, in contradistinction to earlier ideas that they started in Asia. This latter view must be rejected for several reasons. One is that there are no remains of them either in the Late Eocene or in the Oligocene. By the latter, they should have migrated into North America. This may be inferred because they were and are great travellers. An important point is that the earliest Asiatic proboscideans apparently derived from ancestral African prototypes. Even if Asia was not their primary center, it soon became a secondary homeland in which took place the main evolution and adaptive radiation of the elephants. The earliest occurrences of closely allied groups, such as moeritheres, barytheres, sirenians and hyracoids are in Africa.

Figure 8.1: Paleogene profile of Gebel Qatrani at Lake Qarun indicating a significant succession of fossil mammals.

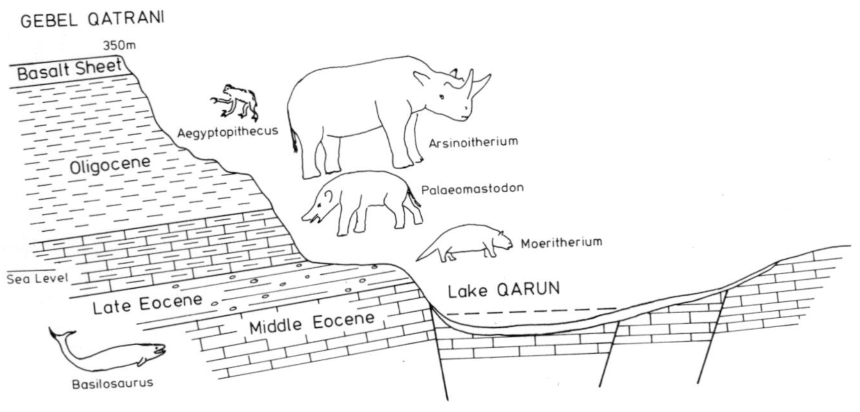

The most primitive true proboscideans of the genus Palaeomastodon are found in the Late Eocene of North Africa. Here also can be traced the origin of all families within the order with the exception of the stegodonts which are the earliest representatives outside Africa. They appeared in the latest Early Miocene after the expansion of gomphotheres and probably also mammutids into Eurasia (2).
After proboscids originated, they rapidly diversi-

fied in Africa into two main adaptive groups. These were the gomphotheres and the mastodonts, known as mammutids in Africa. In the Oligocene, there is practically no record of them. Nevertheless, they must have undergone major radiation and dispersion because, by the Miocene, they have been found in a diverse assemblage ranging from Africa through Eurasia into North America. During the Late Cenozoic, the order underwent an extremely rapid radiation producing new adaptive types destined to dominate most large herbivore faunas of the world. Curiously, in the Late Pleistocene, a dramatic decline changed their fate. Hence, only two species remain on Earth as relicts of the earlier dominance of the group.

In Egypt, fossils of Palaeomastodon have been excavated from the Early Oligocene Qatrani Formation in Fayum, where they are restricted to the Lower Fossil Wood Zone or the Lower Fossil Beds. The former is so named because of its high content of silicified wood, fruits and seedpods together with many fragments of turtles, crocodiles and mammals. The name of this Formation is derived from the Gebel el Qatran, the "tar hills", so named from a black cap rock representing an ancient lava flow which spread over the region after the deposition of the fossiliferous sediments. Palaeomastodon is the main genus in the Fayum Badlands and includes at least four species distinguished by size and dental differences. Male and female individuals may have been recognized among these specimens. Another genus found at this locality is Phiomia, but defining both genera has been a constant challenge to paleontologists since they were designated in 1901 and 1902, respectively (Figure 8.1).

Both Palaeomastodon and Phiomia sometimes reach the proportions of a modern elephant, but usually were much smaller. Comparison between Phiomia and Palaeomastodon leads to the conclusion that the latter is very poorly known from a morphological standpoint. This is because there is no information regarding either its facial and cranial anatomy or the structure of its tusks. In addition, nothing is known of its forest habits and remains of the genus are extremely rare. Indeed, the only morphologic characteristic of any significance in differentiating it from Phiomia, a lowland and savanna type, is the structure of their respective grinding teeth (3).

Continental Tertiary sedimentation began in the Fayum Oasis 60 Ma ago. The beds include the earliest mammals of this age in Africa. The interesting area for vertebrates is between Gebel Qatrani in the north and Birket Qarun in the south. Here, the Paleogene is optimally exposed. In the Late Eocene, the marine Birket Qarun and the fluvio-marine Qasr el Sagha Formations were deposited. This latter comprises 200 m of interdeltaic and deltaic sediments deposited in the Early Cenozoic Egyptian Gulf between the Tethyan foredeep and the Arabian-Nubian Shield. There are four distinct facies. An arenaceous bioclastic carbonate facies is composed of 1 to 3 m thick beds of glauconitic or limonitic, calcareous quartz sandstones and sandy limestone. Occasionally, partially articulated sirenian or moerithere skeletons occur. Interbedded with this is a gypsiferous and carbonaceous laminated claystone and siltstone facies. This indicated an open to restricted lagoonal origin with plant-rich beds ranging from 1 to 6 m in thickness. They pass laterally into an interbedded claystone, siltstone and quartz sandstone facies. It is composed of 10 to 35 cm cycles which become finer-grained upward and constitutes foresets. This facies reaches thicknesses of up to 35 m and may extend as much as 15 km laterally. It is interpreted as the deposits of a prograding delta front. Cut into and interdigitating with these is a quartz sandstone facies consisting of large-scale trough cross-stratified channels up to 35 m thick and 1 km width. It contains a mixed assemblage of disarticulated continental, marine vertebrates and represents distributary channel deposits. Molluscs preserved in the offshore bar and barrier beach deposits imply well circulated, moderate to high energy, warm normal marine, open coast or bar front to more brackish, estuarine mouth environments. They have a firm to mobile sand-silt substrate usually occurring in water depths less than 20 m.

The terrigenous Qatrani Formation is assigned to the Oligocene. Most of it is gravelly sandstone showing early diagenetic alteration and deriving from point-bar deposits of laterally aggrading, meandering streams. Associated mudstones are overbank flood-plain deposits. Small quantities of limestone and dolomitic marl were laid down in shallow, perhaps ephemeral, flood-plain ponds and contain freshwater ostracods and carophytes. The sandstones include abundant silicified woods indicative of an Indo-Malaysian (Tethyan) wet tropical to subtropical climate. The Qatrani Formation yields ninety-five species of fossil vertebrates and thirty kinds of trace fossils, most of the latter being endichnians. Organisms responsible for the traces comprise annelids, crustaceans, insects, mammals and plants. Most interesting perhaps are the fossilized nest structures and gallery systems of subterranean termites. Underneath all these is 200 m of Middle Eocene comprising the oldest beds in the Fayum and including the lower Wadi Rayan Formation and the upper Ravine Beds. The former consists of limestone with Nummulites gizehensis and the latter of shales, marly limestones with Zeuglodon and fish, of the Gehannam Formation. It is disconformably overlain by an Oligo-Miocene olivine basalt or by Miocene fluvial sands (4).

The Fayum fossil mammals are assigned to eleven orders and these were very significant in mammalian evolution in the Ethiopian zoogeographic realm during the Early Tertiary. Among Oligocene forms were huge land-tortoises, crocodiles, gavials and early relatives of the elephant such as the hog-sized Moeritherium and the larger Palaeomastodon (5).

Although Moeritherium greatly differs from living elephants, it had very definite proboscidean orientation and constitutes the oldest member of the order. Small in size, it had still a rather long skull with the eyes set far forward and a nasal opening somewhat to the top. The developing trunk most probably resembled a flexible, tapir-like snout. Tusks had begun to form because the second incisors were greatly enlarged, the upper pointing down and the lower projecting forward to meet them. A diastema was developing so that the first premolars, the lower canine and the lateral lower incisor had been lost. The molars had low crowns with two crosslophs. The rather primitive limbs were of heavy built and show initiation of specializations in the direction of mastodonts and elephants. In spite of the fact that the skull resembles those of the conies and sirenians, the three groups existed contemporaneously so that any common ancestor must have existed in the Early Eocene. Living in the Fayum at the same time as Moeritherium was Barytherium, another primitive proboscidean only known from a jaw and some fragments. Some peculiarities in the former indicate that perhaps the genus was rather distantly related to proboscideans.

Palaeomastodon was much bigger than Moeritherium and structurally more complex as well. The braincase was shorter and higher with nostrils set far up and back on the skull. The snout was probably long and there were real tusks present. The upper ones were borne by the elongated premaxillae and curved down and out. The shorter lower ones were scoop-shaped at the end of the elongated jaws. There were premolars above and two below, the molars being lophodont with low crown. In Phiomia, the molars were more advanced than those of Moeritherium, having three cross-crests instead of two. This genus was probably ancestral to most of the later mastodonts because both it and them had many accessory cusps. They are grouped as the gomphotheres and invaded Eurasia and North America before the end of the Miocene. Here, a bewildering variety of them lived through the Late Tertiary, but only a handful survived into the Pleistocene.

The proviverine hyenodontid, Masrasector aegypticum, may be a cousin of Anasinopa from the Miocene of East Africa. Also, there are numerous primitive rodents called phiomyids which must have lived in undergrowth together with small mammals resembling the living hyrax, the Biblical cony. Larger relatives of the hyrax included the herbivores Geniohyus and Sagatherium which were preyed on by primitive carnivores like Pterodon. A deer-like animal, Bothriodon, is also found here and in the American Big Badlands of South Dakota.

More exotic animals included the giant Arsinoitherium which is so different from other mammals that it constitutes an order of its own, the Embrithopoda, and has never been found outside the Fayum. This huge horned beast was about the size of a rhinoceros and had graviportal limbs with long, massive humerus and femur, short lower segments and a broad, spreading, five-toed foot. It stood about 1.75 m in height and was about 3 m long. By far its most prominent feature were its two great horns fused at their bases on the nasal bones, although there were also small ones on the frontals. An ossification of the partition between the nostrils assisted in supporting these large structures. The frontal region of the skull implies that the animal possessed movable cropping lips, but the row of teeth was complete and there was no enlargement of the incisors such as is seen in other subungulates. The molars were hysodont, a curious phenomenon at this early date in the Tertiary. The upper ones had heavy protoloph and metaloph, while the lower ones had cross-crests indicative of derivation from a double-V pattern. The animal is quite unique in that nothing whatsoever is known about its ancestry or about its descendants, if indeed it had any. Some aspects of it, in particular the molar pattern, suggest a common origin with the conies. If this is true, the relationship must be a rather distant one because the peculiar attributes of Arsinoitherium clearly took a long time to evolve. As regards its mode of life, its skeletal characteristics demonstrate that it could only move slowly on land. Thus, it may have adopted an amphibian habitat similar to that of the hippopotamus or the great sauropods. The massive size in the proportions of the bones argue in favor of this hypothesis.

The Fayum fossil horizons can be defined by their fossil vertebrate contents. These include biostratigraphic zones with Basilosaurus as well as Moeritherium with Pterosphenus in the lower part of the profile. They are succeeded by Phiomia with Palaeomastodon and Arsinoitherium, Apidium moustafai with Propliopithecus haeckeli and Apidium phiomensis with Aegyptopethicus and Aeolopithecus. The first two constitute the Qasr el Sagha Formation of Bartonian age. It is a mixture of deltaic and overlying transitional marine facies containing fish, reptiles, such as sea snakes, gavials, turtles and mammals. The remaining three make up the Qatrani Formation. In its basal beds occur remains of some of the most spectacular mammals in the world. Among these are skulls or partial skulls and mandibles of the enormous Arsinoitherium, Palaeomastodon and Phiomia. There are also coprolites which are mostly reptilian. No doubt the silicified wood and lignites represent former plush forests which maintained birds, turtles (including Testudo ammon rivalling in size the giants of the Galapagos Islands) and small carnivorous and herbivorous mammals, including primates. This contrasts sharply with the paleomagnetic postulate of a circumglobal desert belt extending from the Western Sahara to India through the Fayum at this time (6).

The fourth zone is relatively poor in fossils, but remains of small vetebrates, mostly mammals, are preserved. The reddish sands and gravels in which they are found suggests a continuation of the tropical and humid conditions of the Lower Fossil Wood Zone below. Finally, the Upper Fossil Wood Zone contains less plant remains than the lower and comprises a number of stream channel deposits. From their soft sands come a variety of Early Oligocene mammals, including insectivores, bats, hyracoids, rodents and primates (7).

Contrary to previous opinion, the Fayum fossil vertebrates were not transported for long distances, but probably lived near to where they died. The idea that silicified wood is a practically infallible index to the occurrence of fossil bones in the Upper and Lower Fossil Wood Zones is repudiated. Coprolites are better indicators. Whether above or below ground, gypsum deposits are incompatible with the occurrence of fossil vertebrates.

Palynologic investigations may shed light upon the diet of the fortunate mammals who roamed this primeval Garden of Eden. Angiosperm pollen grains from the Oligocene beds of Abu Roash and the Mokattam region of Egypt include Chenopodiipollis and Caryophyllidites which continued into the Neogene. Paleoecologically, it is significant that Marsileaceae spores and pollen of Gramineae are also found and imply prairie and savanna conditions at that time. The presence of Pentapollenites is important because it also flourished in Europe. Taxodiaceae and Cupressaceae in swampy forests grew both there and in the Fayum. As regards the Oligocene vegetation in these areas, while there are similarities, there are also differences, e.g. no grasslands then existed in Europe (8).

The first recorded injury of a fossil mammal from the Early Oligocene of Fayum was found in the radio-ulna of a young adult Arsinoitherium. It was indicated by a diagonal break across the right foreleg. The ulna was more affected, splintered and developed a swelling on the shaft probably caused by osteomyelitis. Probably the animal wandered around on dry land near the lake and river where it lived and fell down. Because it was so heavy, it fractured its front right lower leg near the knee joint. As it was young, healing was possible and the radius and ulna ankylosed together along the plane of breakage.

Giant birds lived at Fayum during the Early Oligocene and fragments of huge eggshells have been found. These differ very little from recent ostrich eggs and are assigned to Eremopezus which may or may not be a relative of the ostrich-like Psammornis libycus found in Libya.

The injured mandible has a very swollen ramus and this may have resulted also from osteomyelitis. The individual concerned was a rather young adult and it probably died from an abscess caused by the injury. It is coincidental that both this and the Arsinoitherium with the fractured radio-ulna mentioned above were found less than a meter apart in a fossil bone field covering thousands of km^2 (9).

Auspicious conditions paralleling those in the Garden of the Fayum in Egypt must have been more widespread as is evidenced in the southern Sirte Basin of Libya. Near the southern tip of Gebel Al Haruj, there are basal dolomitic limestones of a Middle to Late Eocene age. Overlying them are 40 m of evaporites, themselves discordantly succeeded by another 40 m of sandstones, siltstones and calcilutites. Here may be seen a rich vertebrate fauna. The multitude of fossils includes proboscideans such as Moeritherium and Phiomia, sirenians, crocodiles, turtles, the large aquatic snake Gigantophis and fish, including sharks, rays, sawfish, catfish and lungfish. It has many similarities to the Fayum fauna in Egypt, but seems to be less diversified (10).

REFERENCES

1. SIMONS, E.L., 1974. The relationships of Aegyptopithecus to other primates. Ann. Geol. Surv. Egypt, 4, 149-156.

2. El-KHASHAB, B., 1979. A brief account on Egyptian Paleogene Proboscidea. Ann. Geol. Surv. Egypt, 9, 245-260.

3. MOUSTAFA, Y.S., 1974.The Oligocene African Proboscidea. Part II. The Phiomia-Palaeomastodon question. Ann. Geol. Surv. Egypt, 4, 417-432.

4. BOWN, T.M., 1982. Ichnofossils and rhizoliths of the nearshore fluvial Jebel Qatrani Formation (Oligocene), Fayum Province, Egypt. Palaeogeogr., Palaeoclimat., Palaeoecol., 40, 255-309.

5. El KHASHAB, B., 1974. Review of the Early Tertiary eutherian faunas of African mammals in Fayum Province, Egypt. Ann. Geol. Surv. Egypt, 4, 95-114.

6. VONDRA, C.F., 1974. Upper Eocene transitional and near-shore marine Qasr el Sagha Formation, Fayum Depression, Egypt. Ann. Geol. Surv. Egypt, 4, 79-94.

7. MOUSTAFA, Y.S., 1974. Critical observations on the occurrence of Fayum fossil vertebrates. Ann. Geol. Surv. Egypt, 4, 41-78.

8. KEDVES, M., 1985. Etudes palynologiques sur les

sédiments Préquaternaires de l'Egypte Oligocène. Rev. Esp. Micropal., 17, 333-346.

9. MOUSTAFA, Y.S, 1974. Studies on palaeopathology and taxonomic problems of some Fayum fossil vertebrates. Part I. The first recorded injury among Fayum fossils. Part II. Palaeo-ornithological taxonomy and the first Fayum fossil egg shells. Part III. An injured mandible of (?) Saghatherium. Ann. Geol. Surv. Egypt, 4, 139-148.

10. WRIGHT, A.W.R., 1980. Palaeogene vertebrate fauna and regressive sediments of Dur at Talhah, Southern Sirt Basin, Libya. In: The Geology of Libya, 1, ed. SALEM, M.J. BUSREWIL, M.T., 309-325, London, UK.

Chapter IV-9

BIRTH OF AN OCEAN

"And ye came near and stood under the mountain: and the mountain burned with fire unto the midst of heaven".
Deuteronomy, 4, 11.

TETHYS OCEAN

The Mediterranean Sea, a major relict of the vast Tethys Ocean, reaches depths of over 3,000 m and was created in a vise comprising colliding continents. It is 3,700 km long and 1,100 km at its greatest width and includes a chain of basins, seamounts, rises and islands as well as escarpments and shelves which all testify to the powers of plate tectonics. Its complex topography is the result of collision, buckling and fracturing between continental plates, a process which began some 44 Ma ago when the African Plate moved north into Iberia which was separated by a fault from Europe.

Four structural zones have been discerned in the eastern Mediterranean with diffuse boundaries. These are not microplates but result from the breakup of the colliding African Plate. A piece of northern Africa split off to form another plate called Apulia, abutting on to both Iberia and Europe. By 9 Ma ago, this multiple plate pile-up had constructed most of the familiar existing features. During their development, the widening Atlantic Ocean pushed Iberia and Africa towards Apulia with the concomitant rise of Sicily. The African Plate was forced beneath Iberia, produced magma and this was tapped by volcanoes such as those of the Lipari islands and Etna. The Appenine Mountains in Italy mark the collision between the Apulian Plate and Iberia, forming most of the Italian peninsula. Some trailing fragments of Iberia became isolated as

Corsica and Sardinia. At the same time, Apulia plowed into Europe and thrust up the huge arc of the Alps. To the south, the Atlas Mountains reared up along the coast of Africa as it collided with Iberia. It is interesting that a similar process is going on today under the Aegean Sea, where, moved north by Africa, oceanic crust slides underneath Aegea, a partly drowned landmass characterized by many islands such as Crete.

Figure 9.1: The northern margins of the Afro-Arabian Plates.

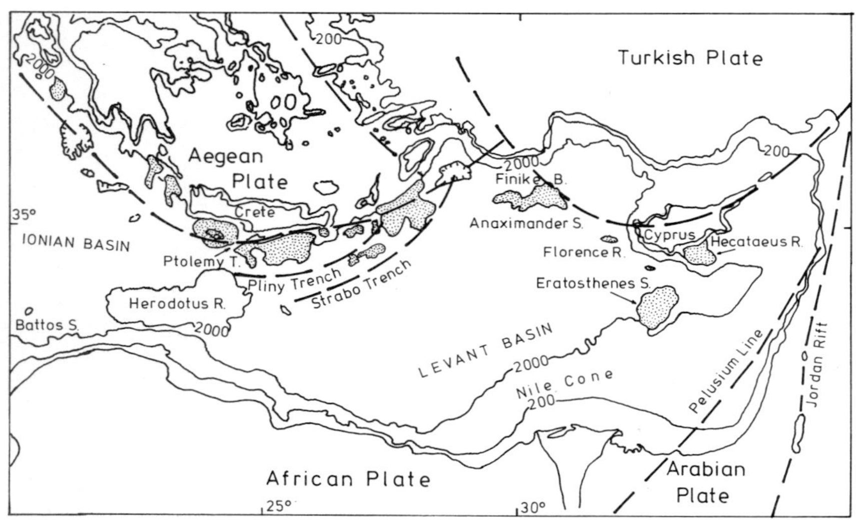

One of the most famous of these is Delos, believed by the ancient Greeks to have been the center of their civilization. In fact, they thought that Delos had wandered through primordial seas awaiting the birth of Apollo and Artemis. This is a remarkable, if fortuitous, anticipation of plate tectonics and drifting. The plunge of Africa both extends and thins the landmass of Aegea, also pulling it away from western Turkey. The Arabian Plate is driving Turkey itself in the same direction and doubtless new mountains will rise when Africa and Aegea collide millions of years in the future and Libya rams Crete. This latter is being rapidly upthrust and parts of it have been elevated 9 m since Roman times. At present, however, the Aegean is one of the few areas in which Africa and Europe have not yet clashed (Figure 9.1).

Side arc collision has been proposed as a mechanism for explaining the crustal extension observed in the Aegean Sea, this describing the interaction of subducted oceanic with continental lithosphere in a subduction arc where oblique subduction occurs. A side arc collision may be occurring in the Hellenic arc near Rhodes. The collision involves subducted African lithosphere, moving to the northeast almost parallel to the arc, with the continental mass of southwestern Turkey. It affects the motion of the Anatolian-Aegean plate complex, but is unlike continental collision since it takes place mainly at depths and involves very little of the shallow and rigid part of the continental lithosphere. Anatolia and the Aegean may comprise one plate complex undergoing counterclockwise rotation. If it were not for the side arc collision near Rhodes, both blocks would show similar deformation and be indistinguishable. At the moment, however, free rotation is only permissible for the Anatolian block, excluding western Anatolia where the motion is accommodated by subduction along the Cyprean Arc. Further west the side arc collision inhibits this rotation along the subduction front. Even further west, undisturbed subduction along the central and western portions of the Hellenic arc is feasible. On the other side of the Anatolian-Aegean plate complex, relatively free motion occurs along the North Anatolian fault zone, including the Aegean Sea. This northern motion combined with the local obstruction near Rhodes creates a torque and a new rotational pattern for the western part of the plate complex and thus a separate Aegean block. The two blocks are not divided by a plate boundary, the Aegean block cannot move freely according to the new torque and effectively moves relatively to Europe and Anatolia by crustal, perhaps lithospheric, extension of which the maximum deformation is seen in the suture zone between the two blocks (1,2).

The Anatolian block is being pushed by the northward moving Arabian plate. It is believed that both interact with the African and Eurasian plates, movements of 5 cm per year for the Arabian and 3 cm per year for the African being assigned. However, the northern convergence of the African and Arabian plates alone cannot explain movements on known faults. An extra driving force is required to produce 2 cm per year westward motion of the Anatolian block. With the addition of such a boundary condition, finite element models give movements along five major faults (North Anatolian, East Anatolian, Dead Sea or Levant fracture, Tüz Gölü and Ecemis) with current magnitudes and directions.

The term "slice tectonics" has been recently proposed and is conceptually different from plate tectonics. Slices are sections of the lithosphere bounded by major geosutures often traversing more than one plate. Also entailed is a global counterclockwise spiralling system of such lithospheric segments. This system may have been imprinted on the lithosphere since the Precambrian, thus having a different origin from the post-Triassic plate tectonics mechanism. Mantle movements induced by the spin of the Earth may have been responsible (3).

The eastern Mediterranean includes a short segment of the convergence boundary between Africa and Eurasia, subduction here being along two very small arcs, the Hellenic and Cyprean. The former is associated with backarc basin and volcanism while the latter is not. Although subduction has been recorded in the Hellenic arc, a continuous trench can not be traced. Instead, there are several deeps of which one, the Strabo trench, is a mere ocean floor cleft whereas its western neighbor, the Pliny trench, is much more prominent. Small oceanic plateaus exist away from the plate boundary.

"Pacific-type" subduction entails a Benioff seismic zone and earthquakes in the Hellenic arc reach depths exceeding 160 km and appear to confirm the existence of such a zone. As here the zone of subcrustal earthquakes is unusually thick (200 km), twice the estimated thickness of the lithosphere in the area, a simple plate model cannot be applied. A similar phenomenon was noted in the northwestern corner of the Cyprean arc, perhaps due to collision and accretion processes. On the other hand, active subduction is going on in the outer trenches such as the Ionian and Strabo ones, but, strangely, Mediterranean lithosphere is only being subducted in the western parts of the Pliny and Ptolemy ones. Inner segments of these last two acted as the main subduction trenches before the subduction zones migrated outward following the collision of oceanic plateaus.

It may be relevant to add that the Eratosthenes shear and associated seamount were named after the "Father of Scientific Geography" who lived from 275 B.C. until he committed suicide in 194 B.C. after going blind and being unable to read. As director of the famous library in Alexandria, a post to which he was appointed by Ptolemy III in 240 B.C., he made his measurement of the circumference of the Earth there and at Aswan.

Figure 9.2: The concept of Eratosthenes

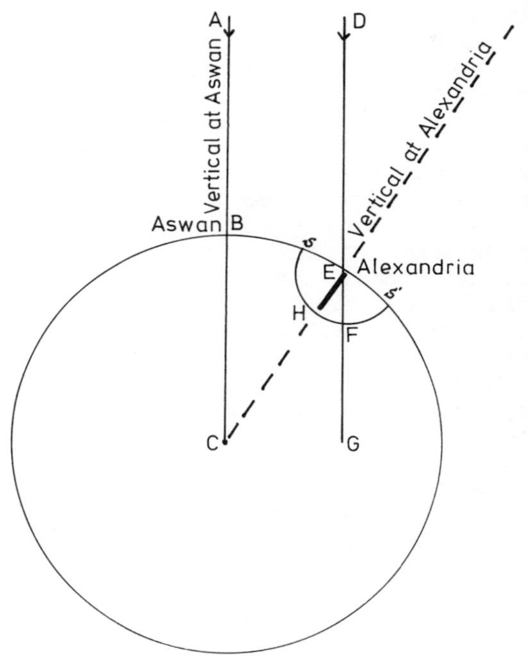

This brought him imperishable fame and was extremely accurate at 46,695 km, too great by 687 km only. This, despite his incorrect belief that Aswan is located exactly on the tropic of Cancer, the northern limit of the apparent journey of the sun, although, for practical purposes, it is close enough. However, it was necessary for him to know only two quantities, i.e. firstly, the zenith-distance of the sun at Alexandria on the day of the summer solstice which he could observe, when no shadows are cast at Aswan, and secondly, the distance from Alexandria to Aswan which he knew approximately.
The actual method used was to employ a bowl, a concave hemisphere from the center of the floor of which projected upward a vertical gnomon (pin) reminiscent of an obelisk. At noon in Alexandria on June 21 or 22, Eratosthenes saw that the gnomon cast

a shadow covering an arc of the bowl equal to one-fiftieth of its circumference. Inferring that no shadow could be cast at Aswan, where the sun is vertically overhead, and as the distance between the two towns was known, this true genius concluded that the circumference of the entire Earth was fifty times this distance taken by him as 5,000 stades as the crow flies. One stade is equivalent to 185.3 m. Thus, 50 X 5,000 = 250,000 stades, which he amended to 252,000 stades, this being the 46,695 km given above (Figure 9.2).

Other famous antique naturalists and geographers included Pliny the Elder, Ptolemy and Strabo after whom three eastern Mediterranean trenches are named. The earliest was Strabo (63 B.C. to 24 A.D.) who wrote a systematic account of Egypt, confining the name to that country which was inhabited and watered by the Nile. Next was Pliny the Elder, born in the year when Strabo died, author of a great compilation of Natural History dedicated to his friend Titus, son of Vespasian, who died investigating the great eruption of Vesuvius in the year of the death of the Emperor Vespasian. Later, Ptolemy, a native of Egypt, wrote two important books, the "Almagest" on astronomy and a geography text. The latter, "Geographike Hyphegesis", fulfilled his principal aim of explaining how to draw accurate maps for which he introduced the concepts of latitude and longitude.

RIFTS IN AFRICA

Africa may have been stationary relative to the mantle since the Miocene. African plate volcanism arising from mantle plumes increased dramatically 25 Ma ago and is still going on, therefore the cause of the activity has not moved relative to the crust. Also, the manifestation of the activity could result from the probability that a stationary plate is more easily penetrated than one which is in motion. The stationary situation of Africa in relation to the plumes may explain the rifting of the continent. In addition paleomagnetic data suggest that there has been no appreciable motion between the magnetic pole and Africa during this time. Nevertheless, this does not preclude longitudinal motion although it does preclude major latitudinal changes (4).

A configuration of horsts and grabens in Libya probably developed when Mesozoic drift of the African plate carried this over a hot spot during Early

Cretaceous. If this existed early enough, it might have split Pangea into its later successors and anyhow could have produced stress capable of fragmenting attenuated and weakened lithosphere.

The East African Rift System is the larger part of a 5,000 km belt of fracturing extending northward from the Limpopo Valley through Ethiopia and the Red Sea to the Jordan Valley. This is a part of the basin-and-swell structure characteristic of the continent. The southern part of the rift faulting in the Limpopo and Zambesi Valleys took place during the Late Mesozoic just before the disruption of Gondwana. It was accompanied by igneous activity involving the formation of plateau basalts, basic dike swarms and alkaline plugs, ring complexes and similar intrusions. The rift fracture pattern may reflect older lineaments of the Paleozoic or even Proterozoic. The northern part of the rift originated in the Middle Tertiary with faulting especially prominent during the Miocene and Pleistocene. There were igneous episodes from the Miocene until today and of course high heat flow, seismicity and volcanicity are still going on.

Structurally, the system is varied. As well as the ideal form of 40 km wide strip with faults on both sides, there are one-sided structures such as the Ethiopian Rift where a normal fault is paired with a monocline, troughs bounded by échélons of faults and regions of irregular block-faulting. Enormous vertical displacements are found, e.g. the floors of lakes in Malawi and Tanzania are 2 km to 3 km below the level of the valley rims with spectacular escarpments at their sides. In some regions, modest thicknesses of Tertiary or Late Mesozoic sediments fill the rift valleys. Mesozoic igneous centers are scattered in the southern part of the rift system, whereas from Tanzania to Ethiopia Tertiary and post-Tertiary igneous rocks are found.

In general, the rift structures tend to avoid those basement provinces in which the tectonic grain is latitudinal, tending to remain within Proterozoic and Early Paleozoic mobile belts enclosing them. In East Africa, two principal branches, the Western and Eastern Rifts skirt the Tanzanian Massif. A similar structural situation occurs in southern Africa, where branches of the system fringe the Zimbabwean and Transvaal Massifs.

An excellent example of intraplate volcanism is that of the East African Rift System, a typical forerunner of continental fragmentation. There is geophysical evidence there to show that, away from the rift valley, there is a normal thickness of crust underlain by an anomalously hot, low density, low seismic velocity body within the upper 50 km of the mantle, i.e. asthenospheric material at shallow depth beneath a lithosphere of which the base was thinned by stoping with the boundary of the asthenosphere during Tertiary to recent times. Domal uplifts in Ethiopia and Kenya represent localities in the rift system where the lithosphere is maximally thinned, therefore representing the focus of the volcanism of the rift. Lateral separation elesewhere is small, probably not exceeding 20 km in Kenya and Ethiopia. Domal uplift also occurs in the Air, Hoggar, Tibesti and Darfur areas as well as in the Adamaoua region of Cameroon, all linked by long wavelength negative Bouguer gravity anomalies. The volcanically active Darfur High is unfractured and links with the meeting place of the Ngaoundere and Abu Gabra rift arms in western Sudan to form an incipient, intraplate, triple junction. These two subsiding rift arms, constituting the Central African Rift System of significantly different character from the East African Rift System, are structurally similar to the incipience of passive margin development, reflecting more closely the inchoation of continental fragmentation than the structures associated with rifting in East Africa. The Ngaoundere and Abu Gabra rift branches may be similar to the debut of the Red Sea and Gulf of Aden Rifts, while the Darfur High completes the triple junction and may be similar to the inauguration of the structures in East Africa (5,6).

This high is associated with a circular negative Bouguer anomaly (50 mGal) 700 km across, the dome having developed contemporaneously with Cenozoic volcanism. It consists of exposed Precambrian rising from 500 to 1,200 m and is superimposed on a broad area of uplift extending from the Chad Basin on the west to the Nile Basin on the east. To the northwest, it includes the Tibesti and Hoggar. The gneisses, schists and intrusives of the dome were mainly folded and metamorphosed during the Pan-African event. Paleozoic granite emplacements and undated pegmatites and dikes cut the NE-SW foliation of the gneisses (7).

The East African Rift System structurally joins a Central African Rift System which began in the Early Cretaceous and is aseismic today. Unlike the former, the latter has not yet developed oceanic-type crust. This suggests that the rift process is developing more slowly than in the other rift system. Alternatively, the regional stress pattern in Africa may have changed during the Tertiary so as to prevent subsequent maturation of the rift system. In fact, the Central African Rift System currently manifests arrested development comparable with that of the East African Rift System in the Miocene. During that Epoch, the sea-floor also started distending.

The Central African Rift System is thought to be an incipient Afro-Arabian type characterized by two distinct rift processes, i.e. extensional tectonics with passive rifting causing subsiding rift branches and active domal uplift. These are associated with quite different gravity anomalies. Extensional tectonic activity is manifested by regional positive gravity anomalies, whereas active domal, laccolithic uplift is connected to a regional negative gravity anomaly.

A further gravity study of 675 Bouguer values for Central Sudan revealed a series of linked gravity minima of 40 km widths and amplitude 200-300 g.u. (20-30 mGal), these anomalies being interpreted as delineating a low density sedimentary infill of a fault-bounded basin called the White Nile Rift. There is a broader 150 km wide positive Bouguer anomaly of amplitude under 300 g.u. centered over the rift. This is interpreted as a thinned crust demonstrating subsidence in the other areas. The White Nile rift is tectonically similar to the southern Sudan rift and the Blue Nile and Atbara fault-controlled basins in that these Cretaceous-Tertiary structures follow similar trends terminating in line at their northwestern ends. Probably this is an echo of a lineament which may extend the Central African shear zone through the Sudan. The form of this is as yet unknown, but it was possibly a major control on the development of deep sedimentary basins in Sudan since Cretaceous times and also on the development of the Red Sea.

Teleseismic P-wave travel time residuals, i.e. delay times, for thirty eight seismograph stations in Africa, these using over a hundred presumed Russian underground nuclear explosions, were established in the middle 1970s. Such residuals seemed to be linearly related to elevations and Bouguer anomalies in the case of stations located on Precambrian crust in the interior of the continent. A resulting lithospheric thickness model indicated a major zone either of thin lithosphere or of thick asthenosphere associated with the East African Rift System and its extension into southern Africa. Concomitant volcanism and seismicity imply that incipient splitting of the enormous African Plate is occurring along the axis of this zone. Unfortunately, no such inference can yet be drawn in the case of the Central African Rift System because insufficient data are available.

A NASCENT OCEAN

The Red Sea Graben started in an Early Tertiary peneplane which covered much of Gondwana. Subsequent dissection and subsidence took place in several phases, ranging from Miocene to Pleistocene, also associated horsts underwent isostatic uplift and tilting. Afro-Arabia was still a coherent unit in the Miocene, at least in Yemen and Ethiopia. When they separated in the Pliocene, northern Sinai and northern Egypt constituted a landlocked isthmus, a continental bridge which still exists. Throughout history, this has been the route of invasion and counter-invasion from Palestine and the Fertile Crescent as well as the door to Asia for higher primates, Homo erectus and later hominids as well as many other animals. Perhaps the best known events were the Flight of the Israelites out of Egypt under Moses and the Flight into Egypt of the Baby Jesus and His parents. However, another major historical crossing was that of Amr Ibn Al-Aas and his Arab army proceeding from El-Arish to Alexandra in the seventh century. With the help of some indigenous Christian Copts in Egypt, he defeated the forces of the Byzantine Empire. It is said that when this commander approached the Nile, he was informed by the Copts that the river has strange habits. On inquiring, he discovered that a custom of the local people was to dress a virgin in her best clothes on a certain night of the year and throw her into the Nile in order to prevent it becoming angry. The story goes that Ibn Al-Aas ordered this to be stopped and instead threw a card in which said "O River, if you flow on your own and have peculiar habits, stop flowing; if it is God who makes you flow, we ask God to keep you flowing". As may be seen, the Nile kept flowing.

At the end of the Miocene, the Red Sea Graben underwent a short period of existence as an enclosed evaporitic basin. Regarding the Red Sea, its axial graben-in-graben structure led to the concept of drift between Africa and Arabia, but its Pleistocene age implies an unbelievable drifting rate of 0.5 m per year. It is possible to model the magnetic anomaly pattern of the Red Sea as a continuous system of sea-floor spreading from the Early Miocene to present by utilizing a time-varying process filter, P(k). Since initial rifting occurred, the half spreading rate is approximately 1 cm annually. Applying such a process filter provides a stimulating approach to analyzing magnetic anomalies over passive continental margins. However, it should be used with caution. Interpretation of the anomaly pattern suggests that the main trough basement comprises thick volcanic sills and flows which were accumulated during early rifting. The extrusions took place over a wide strain zone distributed above the axis of the rift.

A possible Late Cretaceous Proto-Red Sea, an arm of the Tethys Ocean, has been identified from marine, littoral carbonates of the Usfan Formation, containing abundant pelecypods, gastropods, shark teeth and microfossils of Maastrichtian age, these ranging into the Eocene, found at localities in coastal areas of Sudan and Saudi Arabia. The glauconitic sandstone from the middle part of the Usfan Formation gave K-Ar ages of 55.2 to 42.8 Ma and overlies phosphatic beds together with shales, siltstones, sandstones and a massive algal limestone. Paleogeographically, the Mesozoic-Cenozoic shorelines were bounded to the east on the Arabian Shield by the updomed Hail Arch separating two extended sedimentary basins of which the western one stretched as far south as 21° S latitude (Taif, Jeddah). On the western side, the Tethyan arm is traceable as far south as Port Sudan. Similar deposits of this age occur in the Kharga and Dakhla Oases reflecting the shifting Tethyan shores and the associated fossil fauna (8).

The magnetic anomaly pattern over the Red Sea can be used as a paradigm for continuous rifting from the Early Miocene till now with a spreading rate of 2 cm annually since initial rifting. The interpretation of this pattern implies that basement within the main trough of the Red Sea comprises a thick accumulation of volcanic sills and flows extruded during incipient rifting over a wide strain zone distributed about the rift axis. The development of this thickened igneous crustal zone is the second of three successive stages of extension for the Red Sea Basin. The first took place during the Oligocene with convective upwelling. Such an asthenospheric contingency promoted a thermally-induced increase in volume leading to uplift. Then, block faulting and normal fault propagation started in the lithosphere when upwarping and crustal stretching or listric faulting may have happened. Igneous and hydrothermal activity may have accompanied these crustal phenomena and included episodes of volcanic effusion coupled with the emplacement of basaltic sills. Subsequent erosion and non-marine deposition took place. The lithosphere can thin perhaps by divergent convective or diffusive flow in the asthenosphere. Continued thinning and rifting causes subsidence, horsts, grabens and the tilting of blocks. Finally, an extensive marine incursion and the accumulation of thick sediments triggered sea-floor spreading and basaltic ejections with hydrothermal convection. From the Early Miocene to the present, the spreading center narrowed. Finally, south of the 22nd parallel regular sea-floor spreading went on from the Pliocene until now (9).

Paleomagnetic and radioisotopic data confirm that two phases of igneous activity took place in the Late Mesozoic and Cenozoic history of the Eastern Desert. The Late Cretaceous event was more extensive and included trachytes, the Wadi Natash Volcanics and a number of ring dikes together with a regional uplift. The Oligo-Miocene event is evidenced by two basalt flows near Quseir on the Red Sea and by several outcrops in the Western Desert such as Abu Roash. At Quseir, where trachytes gave K-Ar ages ranging from 91 Ma to 68 Ma, the paleopole position is inferred as 63° N, 107° W. This may be compared with 75° N, 131° W, obtained from Wadi Natash intrusions of K-Ar ages ranging between 86 Ma to 78 Ma. The Egyptian Tertiary basalt samples gave paleopole positions varying from 55° N, 131° E, to 73° N, 52° E. From these measurements, it is apparent that Africa was relatively stationary during the Mesozoic and, therefore, it is not feasible to infer stratigraphic periods from paleomagnetic evidence in that part of the world (10).

Both igneous episodes can be related to the tectonic evolution of the Red Sea as also can the regional updoming with its accompanying block faulting and formation of a rift valley at the surface. During this, an enhanced heat flow caused deep crustal metamorphism resulting in dehydration, increased density and alkaline magmatism.

Volcanic fragments from the bottom of the Red Sea, especially from the axial trough show that a flooring of olivine-tholeiitic basalts without modal pyroxene occurs. The relevant magma ascended to reach the bed of the sea before any substantial differentiation took place. Beyond the axial trough occur the above-mentioned basalts with modal pyroxene as well as alkali-basalts, basaltic tuffs and hyaloclastites. The degree of crystallization and fractionation of the melt directly correlates with a mantle plume compensated for by normal faulting and rifting of the axial trough. Thinning and separation of continental crust is limited to this trough. In East Fezzan, Libya, three volcanic phases took place since the Oligocene. During these, basalts were extruded and the most recent are of post-pluvial Holocene age.

East African margins from 15° lat. S into the Gulf of Aden comprise four segments, one of which was produced by the opening of the Gulf as a result of rifting and slightly oblique drifting of Africa and Arabia during the Neogene. Sea-floor spreading is older and more advanced in the Gulf of Aden than in the Red Sea. Arching, orogenic folding and shield volcanism all refer to episodic oceanic crustal construction in these areas. These dynamic effects originate primarily along the rift.

Both the Gulf of Aden and the Red Sea, areally insignificant as they are, constitute an extraordinarily interesting region because they are linked with the typically oceanic Carlsberg Ridge and the continental Rift Valley system of East Africa. Additionally, they are both embryonic oceans in which sea-floor spreading has only recently started. The Carlsberg Ridge itself approaches from the Indian Ocean and is diverted into the Gulf by many transform faults, thereafter continuing as a seismic zone. This latter together with a parallel structure in the southern Red Sea meet the northern extremity of the Ethiopian Rift Valley in the Afar triangle (11).

Several basins extend inland from the coasts on both sides of the Gulf of Aden and are orientated at roughly right angles to the spreading direction, intersecting the coast where sheared and rifted continental margins meet. They look like grabens, one wall of which is continuous with the half-graben of the adjacent rifted margin and it is suggested that they were once parts of a number of discrete rifts arranged en échelon along a zone of lithospheric weakness when the Gulf of Aden started opening. They became redundant when transform faults developed. This kinematic model for the development of rift and transform faults is a new one which may be applicable to the Gulf of Suez (12).

THE RED SEA

The Red Sea is an elongate depression running almost north to south and representing a sub-tropical extension of the Indian Ocean which extends for 2,240 km with average widths of 150 to 325 km and occupies a rift with bordering igneous and metamorphic mountains fringed by sedimentary rocks and active coral reefs. In the northern part, the shorelines are only 180 km apart, whereas southward the gap widens to 360 km. At the south end occur the Farasan and Dahlak Archipelagos rising slightly more than 60 m above sea level and comprising part of a shallow shelf extending almost to the center of the Red Sea. The narrowest point of the latter is reached in the Straits of Bab al Mandab where it connects with the Gulf of Aden. This marks the SW extremity of the fertile part of Yemen, the former Arabia Felix.

It is interesting that the economic decline of Arabia Felix may be attributed, in part, to loss suffered when the frankincense and myrrh market crashed. This resulted from the embracing of Christianity by Constantine the Great in 323 A.D. which led to simpler burials, in turn reducing the demand for these gum resins from the balsam tree and other shrubs. Another cause was the increasing inefficiency, probably due to massive silting, of the Marib Dam in what is now North Yemen. Marib was the capital of the biblical Sheba and the earth-dam was constructed across the Wadi Dhana and faced with stone. Its 600 m length spanned the wadi floor between two sluices and diverted water from flash floods into a canal system irrigating more than 1,600 ha.

North of the straits there is a series of volcanic islands in the middle of the Red Sea which also possesses a distinct main trough from the Zubayr Islands to the southernmost tip of the Sinai Peninsula. Along the present shorelines, there are recent coral reefs built up on biostromal limestones up to 300 m thick. The main trough is cut by an axial trough varying from 30 km width near lat. 20° N to narrow southward where it may be from 5 to 14 km wide. In this axial trough are found hot brine pools and hydrothermal sediments.

The precise time of formation of the Red Sea and surrounding areas is not known. There were premonitory indications in the Late Cretaceous. By the Eocene and Early Oligocene, Ethiopian and Arabian

Figure 9.3: The bathymetry of the Red Sea. The following deeps are shown: Elat (1), Aragonese (2), Dakar (3), Atlantis II (4), Discovery (5) and Suakin (6). Marginal rocks include basemen (crosses) and Phanerozoic sediments (stippled).

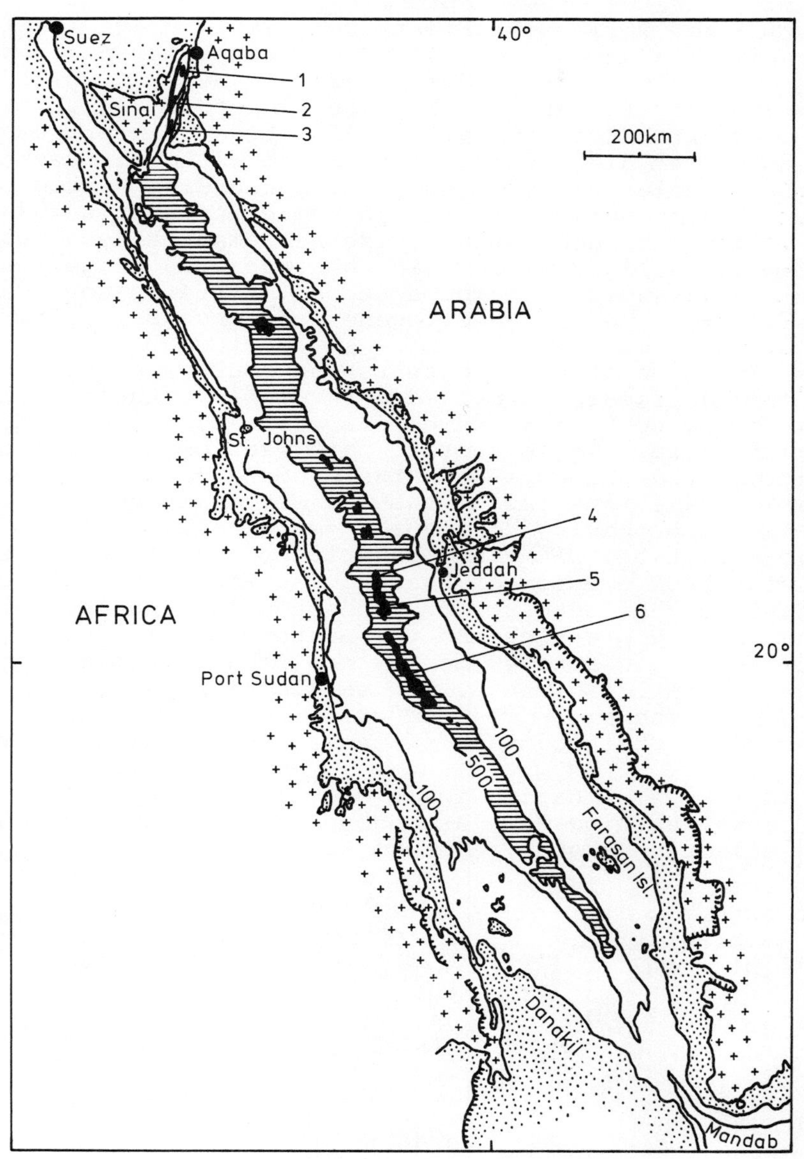

highs over 3,000 and 2,000 m elevation, respectively had resulted from crustal thinning and were separated by the proto-Red Sea. On the summits of these, huge extrusions of alkali-olivine basalts formed plateaus, the so-called Trap Series. In and around the Red Sea depression, there are so few Oligocene marine sediments that it may be that, during this epoch, the Red Sea region occupied the crest of the vast, domed Arabian-Nubian Shield. Monoclinal flexures along the eastern periphery linked with the downwarping of the Red Sea produced a fractured hinge line which filled with differentiating tholeiitic gabbros and diabase dike swarms of Early Miocene age. Although these features suggest rifting, the geological evidence demonstrates that they actually developed in the fractured hinge line mentioned above. In consequence, there is no real indication of pre-Pliocene continuous rifting (Figure 9.3).

The invasion of the Mediterranean southward into the closed depression provided a favorable situation in which an up to 5 km thick evaporite-clastic series could form. The evaporite part went on developing during the Miocene when there could have been no large-scale opening of the Red Sea, although subsidence probably continued. Marginally, clastics interdigitated with the evaporites, the former deriving from erosion of the Arabian-Ethiopian highs. Continued Miocene uplift produced block faulting in the Afar Depression and the eastern margin of the Red Sea was characterized by continued depression of the marginal sediments. The erosional retreat of the edges of the highs created existing escarpments. In contrast, the coastal plains in Sudan and southern

Figure 9.4: Latitudinal section through Arabia Petraea showing an asthenospheric high and the intersecting Aqaba Transform Fault.

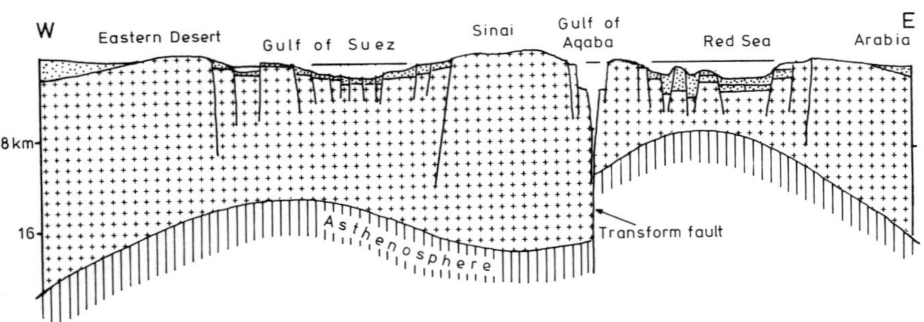

Egypt are not so bordered, the lower slopes merging toward the sea as interdigitating fans and pediments broken by outliers of older rocks and the same is true along the coastal plain of the eastern side of the Red Sea at these latitudes.

To the north, Miocene normal and wrench faulting originated the grabens of the Gulfs of Suez and Aqaba, today the site of treacherous 1 to 2 m tides, which extend southward into the Red Sea depression. Interlayered with the evaporites and clastics, there are localized basalts which show magma generation rather than sea-floor spreading. Seismic information demonstrates that a great deal of the material under the thick Miocene evaporite-clastic sequence could well be Precambrian and not oceanic crust generated at the axial trough in the Miocene (Figure 9.4).

An alkali-olivine volcanism erupted during the Cretaceous and persisted until the Early Miocene. This took place on a continental crust at a converging plate margin. Sub-alkaline tholeiites developed during a time interval of 22 Ma and were connected with wide strike-slip fault movements. This initiated the development of the Red Sea with its associated rifting. A gradual change in the chemical properties of the Tertiary basalts, from primitive tholeiitic type in the Cairo-Suez Road area to more differentiated alkali-olivine type around the Bahariya Oasis indicates advancement from an oceanic to a continental realm (13).

During the Early Pliocene, the Red Sea was incorporated with the world rift system, Arabia breaking away from Africa and opening the straits of Bab al Mandab to permit entry of the Indian Ocean into the Red Sea depression from the south. The Mediterranean could not enter because of the Early Pliocene uplift of the Isthmus of Suez. The port of Suez, or Arsinoe as the ancients called it, developed rather late in history because its harbor was simply a shallow anchorage within breaks in fringing coral reefs. Here, there is a famous fortress, Clysma, with 19th Dynasty remains.

The commencement of drift is marked by the unconformity between the evaporites and overlying marine oozes. Axial rifting in the main trough stimulated the creation of new oceanic crust comprising sub-alkaline basalts, gabbros and diabases. Strong positive Bouguer gravity anomalies and seismic velocities of 6.6 to 7.0 km/sec within the existing axial trough prove that, from the Pliocene until now, this feature resulted from volcanism along an active spreading center with accompanying continued uplift along the Arabian high and extrusion of alkali-olivine basalt.

Recent gravity measurements demonstrate an increase in Bouguer gravity anomaly values towards the Red Sea coast in contrast to regional negative anomalies previously reported for the East African Rift System. In both areas, the anomalies are consistent with thinning of the lithosphere and crustal extension, the difference in the sign of the gravity anomalies resulting from the much greater development of the dilation in the northern Red Sea relative to the rift system. Large negative Bouguer gravity anomalies coincide with granite batholiths outcropping in the Precambrian crust of eastern Egypt. The size of these anomalies indicates that the batholiths are enormous in vertical extent, probably having thicknesses similar to that of the upper crust. Gravity information, therefore, provides structural data on both a local and regional scale. Since the above-mentioned granites are economically mineralized with tin, gold, tungsten and other metals, the gravity measurements constitute a potential tool for prospecting (14).

The Gulf of Suez is mainly characterized by major faulting creating a mosaic of horsts and grabens. This is occasionally modified by strike-slip movements of different trend patterns and intensities. The evidence for this is given by gravity anomaly measurements coupled with drill hole data. Factors producing anomalies include variations in basement relief and lithology as well as basalt sheet occurrences. The depths to basement range from 2 km in the north to over 6 km in the central and southern parts of the Gulf (15).

A borehole in the Massawa Channel entered at 1,400 m depth a sequence of Late Miocene to Early Pliocene submarine basaltic flows at least 600 m thick. All have a composition of transitional olivine-tholeiites and they appear to be intermediate between typical low-K oceanic tholeiites, e.g. those of the Red Sea central trough, and transitional olivine-tholeiites of the spreading axis of the adjacent Afar rift. Problematic minimal ages of 5.8 and 8.2 Ma were obtained from two altered samples. These Secca Fawn basalts extruded either from a localized spreading center in the Early Pliocene or as a result of extensional tectonics on the attenuated continental lithosphere of the Red Sea margin.

Coincidentally with the growth of this sea, the Afar Depression suffered volcanism and its vertical faults remain active. However, there has been no drifting or development of oceanic crust paralleling that in the Red Sea axial trough. Negative Bouguer gravity anomalies across the Danakil Basin and a spectrum of differentiated lavas imply that the Afar Depression is a northern termination of the East African Rift System rather than a continuation of the oceanic rifts. The contemporary seismicity is confined to the axial trough of the Red Sea and fault-plane solutions suggest strike-slip movement along transform faults in a northeasterly direction for the Arabian Plate.

Until the early part of this decade, in the northern Red Sea, the plate boundary between Africa and the Arabian Plate was poorly defined, especially north of latitude 21° N where the trough is less clear and very few earthquakes had been recorded by the World Wide Standardized Seismograph Net. Microearthquake monitoring involves events of less than 3 on the Richter scale and many have been recorded in the northern Red Sea region, but all were to small to be detected by it. A high rate of microearthquake activity has been noted at the southern end of the Gulf of Suez and the hypocenters, of which the depths range from 6 to 14 km, are spatially tied in with major normal faults in the Gulf. Scattered earthquake epicenters define the plate margin south from the Gulf of Suez through the middle of the Red Sea down to about 26° N (16,17).

Between 21° and 26° N in the Red Sea, no seismicity had been recorded, but a very high rate of activity was registered near Abu Dabbab which is about 20 km inland in the Precambrian Red Sea Hills of eastern Egypt at approximately 25° N. This high microseismicity seems to mark the termination of a zone of activity transverse to the Red Sea. However, the extremely tight clustering of hypocenters and lack of a coherent focal mechanism lead to the conclusion that the seismicity is not simply a result of regional tectonics. There are no surface data from which a source may be found. The Arabic name means "Father of the knocking sound" and a nearby locality, Abu Madafi, translates as "Father of the cannon", these suggesting that activity has been going on for quite a long time. One explanation is that it relates to the intrusion of a pluton into the Precambrian crust, the rifted margin of the Red Sea comprising an appropriate environment for such igneous activity. This appears to be very unlikely in view of the fact that the Red Sea periphery is actually passive.

By 1986, several regional earthquakes had been recognized here and at least five events had epicenters in the northern Red Sea. Four of these had epicenters approximately 100 to 160 km north to northeast of the center of the Abu Dabbab seismic zone. A northern Red Sea event was detected during a Gulf of Suez array recording and had an epicenter at roughly 27° N and 35° E. This new microearthquake information demonstrate that the northern Red Sea is not aseismic, even though most of the seismic activity is rather localized. If the relevant epicenters are correlated with the bathymetry of this area, then five poorly located events tentatively define an active tectonic zone which extends from the southern end of the Gulf of Suez into the median zone of the Red Sea, following the deep water axis and the spreading axis implied by magnetic anomaly profile data. This zone of seismicity may be an active spreading zone connected with the opening of the Red Sea. The high degree of microseismicity at the southern end of the Gulf of Suez is linked with the triple junction of the African and Arabian Plates and the Sinai Peninsula. In antique times, the last area formed the southern part of Arabia Petraea.

The observed events seem to derive from fault planes roughly parallel to the Gulf of Suez, but they must be coordinating adjustment in motions among different rates of oblique opening in the Red Sea and Gulf of Suez as well as the left-lateral strike-slip motion in the Gulf of Aqaba. On the eastern side of the latter, the results of compressional tectonics are apparent. They are unusual and explicable by inferring a former collision between Midyan and Sinai. In fact, several collisions took place at an early stage due to a curved strike-slip fault system in the northern Red Sea (18).

Despite the recording of over 200 microearthquakes in a very small area near this locality so far it has been impossible to correlate this with active faults in the region. However, it implies a displacement in the activity of the plate margin perpendicular to the trend of the Red Sea. The combined geophysical and geological evidence clearly shows that the Red Sea axial trough is a rift in which new crust is forming and that rifting began in the Pliocene, although other features originated in the Late Cretaceous. These latter cannot be conjoined to the existing world rift system.

HOT BRINES

The Red Sea and the Gulf of Aden belong to an ocean still in parturition. The central trough of the Red Sea shows gravity and magnetic anomalies as well as heat flow and seismicity. Tholeiitic basalt is ascending from the upper mantle. Sea-floor spreading is going on at a rate of 1 to 2 cm annually coupled with thinning of the crust. During the Oligocene and Miocene, a graben with peripheral volcanism developed in the Arabian-Nubian Shield. Here, the incipient Red Sea subsided with accumulation of over 5 km of clastic sediments and evaporites. There was a connection with the Indian Ocean since the Pliocene.

Also, precipitation of iron and manganese in the Miocene indicates that hydrothermal activity had started. There were ore solutions 25,000 years ago in the Atlantis II-Deep, but concentrated heavy metal deposits only began forming 12,000 years ago. Brines of the central trough and the deeps are associated with seawater descending along faults which dissolved Miocene evaporites and leached metals from the sediments. Circulating brines ascend along the axis of the rift because of increased heat flow. Different deeps contain dissimilar hydrothermal systems. Isotopic data are inconclusive, the lead, helium and strontium isotope ratios suggesting an upper mantle provenance, whereas the sulfur isotopes are rather ambiguous (19).

The above information permits the division of the deposits into normal, hydrothermally influenced and metalliferous sediments. Samples from deeps in the Red Sea show three facies groups on the basis of their clay mineral content after carbonate dissolution, namely, normal sediments, normal sediments with hydrothermal influence and heavy metal deposits. As regards such sediments in the Commission Plain, radiocarbon model ages derived in connection with micropaleontological investigations vary from 6,600 to 2,800 years for the surface sediments. This results from reworking, residimentation and input of allochthonous material. The depositional rates average 4 to 5 cm per century. The sulfide facies at the end of the Pleistocene ranged from 11,600 to 10,600 years B.P. and records hydrothermal activity with depositional rates of 5 cm per century. The underlying manganite and limonite facies contain hard layers of aragonite and give ages between 14,800 and 11,650 years B.P. In this area, the maximum age of the sediments is 19,700 years. Transport from the east is indicated by a general thickening in that direction.

DSDP leg 23 sediment and rock samples from the central trough area of the Red Sea enabled several lithologic units to be identified. These are a basal evaporite sequence of Late Miocene age or older having a sabkha origin, a unit with high organic carbon content and up to 10% pyrite and overlying marine Plio-Pleistocene beds reflecting the opening of the Red Sea to the Indian Ocean. Dark shale layers in the lower part of the last beds are enriched in vanadium and molybdenum and are comparable to the Kupferschiefer. The high organic carbon and pyrite content of the Pliocene-aged dark layers imply deposition under euxinic conditions.

The concentrations and isotopic compositions of strontium in the non-carbonate fractions of sediment from a piston core taken from the median valley of the Red Sea demonstrate that the silicate fraction of the sediment is a two-component mixture. This results from the mixing of volcanic-basaltic and terrigenous-sialic constituents. The former are weathering products and include volcanic dust while the latter also result from weathering of old sialic rocks in the neighboring land masses of northeastern Africa and the Arabian Peninsula. The average Sr-concentration of Holocene volcanics from the Red Sea islands is 774 ppm and the Sr-isotope ratios of basalts of this age average 0.704. Calculations of the concentrations of the basaltic components in the sediment were utilized in order to define varying horizons and this facilitated refinement of the conventional lithologic stratigraphy.

A specialized case of ridge crest accumulation is the heavy metal deposits of the Red Sea, the most important of which occur in the Atlantis II-Deep off-shore of Mecca. This is more than 2,000 m deep and only 6 by 15 km in extent, the bottom being filled with a hot brine at 60° C and a salinity of 25% which is seven times that of seawater. It is a hydrothermal sedimentary deposit still being formed and comprising a trap caused by northwesterly faults limiting local grabens. These conditions have existed at least since the end of the Pleistocene, but were frequently disturbed by lava flows, tectonic events and displacements of the discharge vents.

As a result, varied lithologic facies arose with a depositional differentiation from sulfidic-siliceous precipitation near discharge areas to iron and manganese oxides formation outside the brines. The

recent hydrography and the chemistry of the brines in the Red Sea suggest that iron hydroxide is formed largely by the oxidation of initially precipitated manganese hydroxides. Two brines occur, a lower one at 60°C (top 2,044 m) and an upper at 49°C.
During the late 1960s, the hot brines of the Red Sea received 0.346 km^3 of water with a minimum temperature of 104°C, this probably originating from a relatively shallow depth. Since the temperature of the brine has increased, the process forming the underlying deposits, rich in heavy metals, is still going on. Metal concentrations are anomalously high, the Fe content being a factor of 8,000 times richer than in seawater, the Zn 500 times and the Cu 100 times. The underlying sediment is multicolored and it originates from two processes. The first is volcanism on the axis of the spreading center which may lead seawater to react with freshly rising basalt. The second is hot water circulation through gypsum and salt layers of adjacent thick Tertiary deposits abutting against newly formed sea-floor. This leaches out metals and salts which produce the metalliferous brine collecting in the brine pools. The metals precipitate on cooling as well as when oxygen is supplied by mixing with normal seawater. Such mixing is hindered by the high density of the brine and can take place only near its surface. Consequently, the metals become trapped in the brine. As regards the ore potential of the deposits, in the Atlantic II-Deep alone there are said to be 3,200,000 tons of zinc, 800,000 tons of copper, 80,000 tons of lead, 4,500 tons of silver and 45 tons of gold (20).
Varying morphology and tectonics stem from relative motion between Arabia and Africa along the Red Sea. A deep axial trough bisects the main trough south of 21° N and formed by sea-floor spreading during the past 4 Ma. It is associated with large amplitude magnetic anomalies and high heat flow.
High heat flow in the Red Sea is linked to the spreading center and new data from Egypt indicate that an anomaly extends about 30 km inland from the coast in the Precambrian rocks fringing the Red Sea. Preliminary heat flow values ranging from 1.0 to 4.2 HFU have been estimated for this country, 1 HFU being a millionth of a calorie per cm^2 per second. Heat flow west of the Nile and in northern Egypt is believed to be low, 1.0 to 1.1 HFU, typical of a Precambrian platform province and connected with the low heat province of the eastern Mediterranean. However, east of this river and in the Gulf of Suez,

it is elevated at several sites rising to a maximum of 4.2 HFU measured in a Precambrian granitic gneiss approximately 2 km from the Red Sea coast. Water geochemistry data confirm the high heat flow values, but do not show any deep hot water circulation systems. There is a relatively positive Bouguer gravity anomaly near the Red Sea coast which, together with the heat flow data, implies that the eastern Egyptian heat flow anomaly is connected with the opening of the Red Sea (21).

Incidentally, a vein-type hydrothermal occurrence of radioactivity at Wadi Um Gir in the central Eastern Desert of Egypt is structurally controlled by a major fault. This is one of the youngest related to the formation of the Red Sea during the Tertiary. Anomalously high radioactivity is connected with secondary uranium mineralization. There is a radioactive disequilibrium with a deficiency of daughter nuclides possibly resulting from their being redistributed by groundwater (22).

Heat flow values from the Kenya portion of the East African Rift System, also from the lakes in the western rift and extending south to Zambia, are also high, but all other heat flow data from Africa show a regionally low to normal heat flow. The high heat flow values probably relate closely to the active tectonic sections of the African Plate, the East African Rift System and the widening ambit of the Red Sea.

In this latter, north of 25 N, the main trough floor looks irregularly faulted. In the north, the magnetic field is characterized by smooth low amplitude anomalies, but there is a few isolated higher amplitude magnetic anomalies usually associated with gravity anomalies and often these are due to intrusions. Between these latitudes, there is a series of deeps irregularly distributed along the floor of the trough and these are marked by large amplitude magnetic anomalies.

The different regions may represent successive phases in the rifting of a continent and the development of a continental margin. At first, there could have been a period of locally diffuse extension by rotational faulting and injection of dikes. This may have been followed by a linear concentration of extension along a single axis and the start of sea-floor spreading. during the Miocene, the main trough of the southern Red Sea formed by the same process of diffuse extension which is still active in the northern Red Sea. This model would explain available data as well as reconciling

previous models which emphasize either the evidence for considerable motion between Arabia and Africa or the evidence for downfaulted continental crust beneath much of the Red Sea (23).
The Red Sea floor descends in two steps to the Suakin Graben. The western one is occupied by the Suakin Deep-Sea Plain in which there is a high rate of sedimentation of marls of 4 cm per century, but no significant evidence of hydrothermal activity. Heavy metals occur in the sediments and concord with normal Red Sea sediments. To the east, the bottom of the Suakin Graben comprises a thin sedimentary cover with local outcrops of basalt. At the maximum depths of the graben, brines occur in two separate basins in which the sediments are different.

THE RIFT IN ARABIA PETRAEA

In the Dead Sea and Jordan Valley, there is a major lineament demonstrating relative movement between Arabia and the Mediterranean, the post-Cretaceous displacement on this being 105 km which implies a rate of dislocation of 0.4 cm annually since the Late Miocene. Tensional stresses in the Red Sea in the Pleistogene produced similar kinematic effects. The regional transform fault of the Rift System is a strike-slip feature, manifesting itself in a narrow depression that extends from the Gulf of Aqaba or Elat for approximately 360 km to Lake Tiberias and separates the Arabian Shield from the Levant. Unconsolidated sediments of Neogene and Pleistogene age occupy most of the Rift floor.
About 60 km off Israel, a questionable lineament, the so-called "Pelusium Line" could be the western boundary of a postulated "Central Plate" incorporating Sinai and wedged since the Paleozoic. It may extend to the Central Sahara and be represented by a major fault line of the same trend. This is taken to be a major subcrustal fracture and the chain of Abu Roash, Fayum, Bahariya and Farafra structures is related to this fault as well. Evidence for this is provided by positive anomalies of gravity (24).
The Gulf of Aqaba is structurally a northern continuation of the Red Sea Rift where, under arid conditions, carbonates are accumulating together with clastics. The latter are deposited mainly in alluvial fan complexes spilling out on to a narrow shelf. The carbonate deposits include reef complexes

and lie along the shelf. Calcium carbonate cementation of submerged carbonated reefs has reduced their porosity to 28% and their permeability to 0.01 m/d. However, adjacent subaerial Pleistocene carbonates have undergone leaching and the development of secondary porosity through dissolution by meteoric fresh water. In them, the porosity may attain 60% and the permeability 10,000 m/d. From this Gulf, the Rift floor rises gradually to elevations of 250 m in the central Wadi Araba. Thereafter, there is a steady decline northward to the Dead Sea almost 400 m below sea-level.

The northern Dead Sea contains a trough which is 400 m deep and may have had an atectonic origin. This means that it formed not by fault movements, but perhaps by solution of a salt body during the last 15,000 years. The various tectonic causes proposed cannot explain why well-exposed layers of a former Lisan Lake and its shrinkage terraces, of which the age cannot be much more than 15,000 years, have maintained their undisturbed and semi-horizontal position. For instance, if transverse faulting actually occurred within this time, downthrows of at least 400 m have to be assumed. Such impressive features should continue in the fault pattern of the mountains bordering the Dead Sea Graben and ought to be found in other parts of the Jordan Graben as well as in other areas of the countries of the Levant. However, they are not.

At all events, gravity profiles across the Dead Sea show a correlation between the sign and magnitude of the Bouguer anomaly field and the width of the rift axis. About 100 km is the critical distance for the Bouguer gravity anomaly to change from negative to positive. The inference is that this degree of separation must occur before denser material can rise from below and offset the negative effect of the shallow, thick, low density sediments typically present in continental rift zones. Large scale earthquakes occurred in the past and minor temblors today confirm recent tectonic activity between the Mediterranean and the Dead Sea Rift (25).

The average salinity of the Dead Sea water is 31.5 % and the sulfate-ion concentration is extremely low, but that of bromine, at 6,000 ppm, is possibly the highest occurring in any surface water in the world. The Dead Sea is not a relict body of seawater. Its salts accumulated in a closed inland basin under conditions of aridity and derive from two sources, two-thirds from saline springs and one-third from the Jordan River. The age of this sea could be anything between 70,000 and 12,000 years, the latter being more probable. The annual amount of chemical precipitation in the southern part is 0.3 g/cm^2 and sodium chloride with calcium sulfate are the predominant components, but there is also a lesser quantity of calcium carbonate.

The vast bromine reserve exceeds 800 million tons and its origin is in dispute. It has been ascribed to recent volcanism, but no known bromine-rich magmatic emissions have ever been recorded. In fact, the Cl:Br ratio in igneous rocks is very low. Since most oil waters contain considerable amounts of this element, an organic origin has been proposed. However, this is due more to total higher salinity than to any specific bromine enrichment and anyhow oil waters high in bromine are also invariably high in iodine. The Br:I ratio in Dead Sea water exceeds 105, thus excluding any organic derivation of the bromine.

Most likely, Dead Sea bromine results from its concentration in fossil residual salt brines. Large lakes certainly existed in the Dead Sea Rift since the Oligocene. The Dead Sea became established in the Late Pleistocene in the area of the preceding Lake Lisan which may have dried up before it formed. This was even more saline than the present Dead Sea as is shown by the enormous rock salt masses of the Sodom Formation.

The Dead Sea is rapidly becoming a mudflat because of the tremendous extent of irrigation which diverts Jordan River water for agriculture and causes a recession of the sea of 0,5 m annually. A conduit has been proposed by Israel which would connect the Mediterranean to the Dead Sea over a 113 km long stretch, three quarters of it through the mountains, and restore the latter to its natural level. The seawater will drive turbines and generate 20% of the peak electricity demand of Israel. An even more ambitious and difficult undertaking would be the proposed conversion of the Qattara Depression in Egypt into a large lake. This was envisaged over half a century ago, but there is no feasibility study available within the financial restraints of the country.

THE TRIPLE JUNCTION AT AFAR

The Afar, the Red Sea and the Gulf of Aden comprise the most accessible and extensive region in the world where rifting and drifting can be studied. Within it, processes such as uplift, block-faulting, volcanism, seismic activity, hydrothermal circulation and sea-floor spreading take place.

Attempts at realigning the Precambrian structural trends in the Arabian-Nubian Shield clearly show an overlap of the Afar Depression with the Yemen and pose a serious difficulty for paleogeographic reconstruction. In addition, it is an impediment to the idea that most of the Red Sea is underlain by freshly formed oceanic crust. Magnetic anomaly data and structural features show that most of the depression is underlain by sima created during the separation of Arabia from Africa. This mitigates the problem by reducing the actual area of overlap in palinspastic reassemblies based on shore-line fit.

There are two possible explanations for the spreading zone of the southern Red Sea. The first involves stretching, unlikely because the known zones of expansion are narrow. The second entails assimilation of the lower part of the continental crust by hot ascending mantle material prior to spreading. Later dilation and lateral motion could have formed the marginal zone by stretching, block faulting and subsequent introduction of new oceanic crust under the axial zone (26).

Along the Red Sea Rift, the crust ruptured completely and released basaltic material of the inner graben. The magnetic anomaly pattern implies that sea-floor spreading is occurring now, also that the intracontinental and mid-oceanic rift systems are essentially similar.

The Afar is framed by the Ethiopean Highland in the west, the Somali Plateau in the south and the Danakil Horst in the east. Its northern part is occupied by the Danakil Depression, a branch of the Red Sea. To the west, the area consists of highly disturbed basement overlain by unfolded Mesozoic and this is true of the Danakil Alps as well. Within the Afar Depression, Neogene sediments lie unconformably on pre-Tertiary formations. In them, older limnic-fluviatile beds are succeeded by marine deposits. The whole sequence thickens towards the center of the depression.

These basin fillings constitute the Danakil Formation and indicate a major phase of rift faulting before and during their accumulation. Their extensions on both flanks of the depression reflect the structural borders of the Danakil Graben. Evaporites occupy its deepest part which was downfaulted or opened by major rift movements in the Pliocene. Quaternary rifting determined the present topography and was succeeded by a marine invasion. Its sediments make up the Zariga Formation and gave radiocarbon ages of 25,000 to 34,000 years. They enclose the depression, passing into gypsum beds toward the center of the basin. The deepest part of the low is concealed under the Afrera Formation with a radiocarbon age of 5,800 years. This frames several lakes.

In the northern Afar, rifting accompanied strong volcanic activity so that extensive basalt flows intercalate and locally underlie the Danakil Formation. The Afar basalts gave K- Ar ages ranging from Miocene to Pliocene. In the southern part of the Danakil Graben, the late Tertiary sedimentary basin fillings are replaced by these plateau-forming basalt flows. They are succeeded by scoriaceous alkali-olivine-basalts of which the differentiated lavas formed huge volcanoes parallel to the rift structures. In fact, the Central Volcanic Range, which delineates the middle part of the Danakil Graben, is still active.

The Afar triangle cannot be considered as the denouement of recent sea-floor spreading on land. The evidence for this is that oceanic magma has not been found outside the Danakil region and also that its Alps are rotated in an counterclockwise direction to a maximum of only 8° . It is concluded that Arabia went north rather than northeast since the Miocene (27).

During the Pliocene and Early Pleistocene, the Red Sea-Gulf of Aden area opened. This event coincided with an uplift of the southern prolongation of the Danakil Horst. At the same time, uplifts and basins developed. In the Middle Pleistocene, the Gulf of Tadjoura distended as a western prolongation of the Aden rift. Since the Early Pleistocene, coastal uplifts can be correlated with successive alluvial deposits.

The decreasing amount of spreading in the southern Red Sea is compensated for by crustal spreading in the Afar where the geophysical properties are intermediate between continental and oceanic as exemplified by Ethiopia and the Red Sea respectively. Crust of this intermediate type may result from a process of "oceanization" of formerly continental crust by fragmentation and basification through massive dike injections due to mantle diapirism. This concept with resultant sea-floor spreading may be compatible with a Cloosian Hebung-Spaltung-Vulkanismus succession. The various stages are uplift, evolution of graben systems and volcanism, oceanization with crustal spreading, evaporites and widening of the sea-floor. All are recognizable along the Afro-Arabian rifts. The depression resembles recent oceanic spreading centers, although, to some degree, it has a structure intermediate between continental and oceanic. There is an axial graben like the median valleys occurring in many mid-oceanic ridges. Lava flow samples from this are all normally magnetized and young, implying that they were emplaced during the Brunhes normal epoch of the magnetic field of the Earth. Older lava samples gave directions just below the horizontal, but with easterly declinations. These directions may be due to a deviation of the terrestrial field from its normal axial dipole direction. Alternatively, they may have been caused by large-scale rotation about a vertical axis. This rotation could be related to the distortion required to close the continental parts of crust around the Gulf of Tadjoura (28).

Plate tectonic analysis of the Afar triple junction gave integrated extension rates for the Ethiopian rift valley of between 1 and 3 mm annually, but the problem is whether the strain accumulation is episodic or progressive. A northern geodimeter network traverses the rift floor just south of where the valley opens out into the Afar Depression, crossing an en échélon offset in the Wonji fault belt at lat. 8° N which comprises the youngest volcanism and faulting in the rift floor. The surveys reveal a progressive rift extension at mean rates of 3 to 5 mm per year and strain rates of 6 to 16 X 10-7 annual scale change. These results are for the Wonji fault belt segments and it is impossible to say whether they are local deformations, regional or plate tectonic phenomena. However, the current aseismicity of the East African Rift implies a build-up of regional strain. On the other hand, in the Afar Depression, where the Ethiopian Rift meets the Red Sea and the Gulf of Aden, the earthquake epicenters concentrate along peripheral block faults, but there are too few fault plane solutions to allow any inferences regarding motions.

ARABIA AND THE ZAGROS

In the Red Sea, there is a 15 km deep crust-mantle boundary which is overlain by 3 to 4 km of mixed sediments. This could reflect either oceanic crust with a thick oceanic layer 2 or a strongly modified continental crust. There is a clear oceanic crustal structure in the southern part of the Red Sea. In the shelf area, there is a sedimentary basin, mainly filled with Miocene salt, underlain by a double-layered Moho of which the upper layer is interpreted as a zone of intrusion in the viscosity minimum of the lower continental crust and embryonic rifting. There is also a zone of high velocities within the Azir Mountains on the Arabian Shield, these being probably uplifted lower crust.

A 1,000 km seismic refraction profile has been obtained by a traverse from Riyadh to the Red Sea. This crossed three major Precambrian tectonic provinces and Cenozoic rocks of the coastal plain near Jizan to terminate at the outer edge of the Farasan Bank in the southern Red Sea. Results showed that the Shield comprises two layers, each of which is 20 km thick, with mean velocities of 6.3 km/sec. and 7.0 km/sec., respectively. West of the Shield-Red Sea margin the crust thins to a total thickness of less than 20 km, beyond which point the Red Sea shelf and coastal plain are believed to be underlain by oceanic crust (29,30,31).

Underneath the Shield, the Moho varies in depth from 43 km with a mantle velocity of 8.2 km/sec. in the northeast to 38 km with a mantle velocity of 8 km/sec. in the southwest near the Shield-Red Sea transition. Two velocity discontinuities were found in the upper mantle at 59 km and 70 km depth. The crustal and upper mantle velocity structure of the Arabian Shield is interpreted as complex crust originated by suturing of island arcs in the Precambrian. It is currently flanked by the active expanding boundary of the Red Sea.

In the Shield, there are some broad latitudinal folds formed in Late Paleogene, the axes of which become more widely separated towards the north. As the shield warped, an accompanying maximum orogenic phase took place in the Taurus mountains together with folding of the foreland belt, i.e. Lebanon and Palmyra mountains. Early Tertiary volcanism occurs in Yemen and adjacent countries. Evidence from the Gulf of Aden suggests that basic magma was emplaced to form a quasi-oceanic crust (sea-floor spreading) in the rift trough which presumably causes displacement of the continental blocks.

The Aden volcanic series comprises six volcanoes along the south coast of Arabia of which lavas range in composition from mildly alkaline basalts to peralkaline rhyolites. High-precision isotope data from these show a correlation between the strontium and the rubidium-87/strontium-86 ratios, this being equivalent to an isochron of 33.1 Ma, indicating an age far greater than the 6 to 5 Ma age of eruption documented by published K-Ar dates. The older age is interpreted as the time of a fractionation event in the source region which was later subjected to large degrees of melting, producing magmas of different compositions.

As regards the Red Sea, magnetic data from there have been interpreted as indicating that spreading took place in two episodes, the first ending 34 Ma and the second recommencing 5 Ma ago. The crustal composition below 4 to 6 km of shelf and coastal plain sediment is obscure because of the inadequate understanding of the actual amount of separation between the plates. A maximum divergence would entail a mafic igneous buried crust derived from the mantle and emplaced during the formation of the nascent ocean. In this case, the term "oceanic" may be applied to it. An oceanic crust model predicts a maximum of 380 km gap between the Arabian and Nubian Shields through Red Sea-floor spreading. This would necessitate a 35 km or so overlap on a palinspastic reconstruction. A continental crust model suggests a 320 km difference with a narrow interval between the shorelines. Unfortunately, both approaches conflict with kinematic analyses from the northern part of the sea which constrain matters on the basis of limits imposed by rifting models in the Gulf of Suez and fault offset on the Dead Sea transform (32).

Whether the 34 Ma old episode is significant for the Gulf of Aden as has been suggested, or whether spreading followed some different pattern in the latter, as magnetic data may indicate, is not yet known. However, submarine topographic evidence confirms that there was certainly more than one episode of spreading in the Gulf of Aden.

The As Sirat plateau volcanic rocks in Saudi Arabia erupted 29 to 24 Ma ago producing initially picritic lava. Then came alkali-olivine basalt, basanite and finally, andesine-bearing alkali-olivine basalt (hawaiite). Related feeder pipes have similar compositions and indicate by their distribution that the original outcrop was much larger than the existing one.

In the Early Miocene, 22 Ma ago, the Gebel at Tirf volcanic rift zone developed over a short period as a result of crustal thinning and warping and was succeeded by Pliocene to Oligocene eruptions of alkali basalt on the tilted zone. The Red Sea axial trough sub-alkaline tholeiite resulting from the continental break-up of Africa and Arabia was contemporaneous with the coastal plain alkali basalt. The strontium isotope ratios show that all the southwestern Saudi Arabian volcanics probably derived from the same mantle material which was later modified by partial melting, various degrees of fractionation or tectonic setting. The alkali basalts of As Sirat, like those of Yemen and Ethiopia, seem to occupy separate, large-scale, northward-orientated arches in the Precambrian basement and to have preceded rifting in the Red Sea by 30 to 20 Ma.

The earliest volcanism of 22 Ma ago related to the Red Sea structure was crustal attenuation and the formation of the Gebel at Tirf volcanic rift zone. This short volcanic event was succeeded by a long-term period of evaporite deposition in the Red Sea depression and only intermittent volcanic activity. Red Sea axial rifting began in the Pliocene with the formation of axial trough subalkaline basalt and apparent continuation of alkali basalt volcanism on the Arabian Plate along pre-existing structural highs.

The Neogene to Pleistogene Aden Volcanics Belt extends from the East African Rift System over the west Arabian Shield as far as Turkey, thus trending roughly parallel to the Red Sea and indicating tectonism. Warping on the northern periphery of the shield, where the Aden Gulf-related structures end, is connected with the rifting in the Red Sea. Faulting along the Aqaba-Dead Sea system is of the same age and cuts the foreland belt. The folding of the Zagros Mountains is also of Miocene age, an effect of the Alpine Orogeny which actually continues until now. During the Neogene, their external zone was regularly folded and overthrust by rocks of their internal zone, displacements of as much as 100 km having been recorded. Most of them take place along transcurrent fault systems which are almost parallel to the northwesterly fold tendency. The Zagros fault follows the major lineament in this direction, both it and the latitudinal North Anatolian fault being subject to dextral displacements at present.

The associated effects are still manifested in the highly seismic regions of Iran and Turkey. In fact, crests of folds in the petroliferous limestone and shale succession of the external zone in the Zagros Mountains were still rising recently as shown by the deformation of Ababassidean aqueducts which must have been built between the 8th and 13th Centuries. Probably the fault systems already mentioned and those of the Jordan Valley act as an adjustment mechanism permitting the displacement of the Arabian Shield to the northwest in response to the oblique opening of the Red Sea.

A quarter of North Yemen is covered by volcanic rocks comprising alternating lava flows, basalts, andesites or trachyte porphyries and various kinds of colored tuffs having a thickness probably exceeding 1,200 m. In this series are intrusive basalts distributed as dikes, sills or laccoliths and the old major volcanic centers are rare. They can be observed either as chimneys or thickenings of lava flows. During this eruptive period, there were quiet intervals during which freshwater or alluvial deposits or sands or paleosols of lateritic type accumulated.

The volcanism started in the south about the end of the Cretaceous, but became more intense and extensive during the Tertiary. Characteristic are extrusive and explosive phases related to tectonic deformation along faults. It continues until now. There are gray-white granite hills protruding above the topography carved out of the volcanics which they deform, thus being younger. Earlier granites include calc-alkaline ones which were emplaced during the Early Paleozoic after the Pan-African Event. This is established by the apparent age of 652 Ma. These post-tectonic rocks include not only this type, but alkaline granites of which some relate to Cretaceous sedimentary and volcanic rocks of the Trap Series, suggesting intrusion through them. Indeed, one instance, the Gebel al Munif granite complex, arose later than contiguous basalt lavas and may be later Tertiary in age. The alkaline granite may result from differentiation of basaltic magma because they are associated with mafic extrusives. If this can be proved, the latest granites could have had a quite different origin from the bulk of Yemeni granites. Some of these intrusions are 10 km long and, just like the still conical volcanoes, they are Holocene. In one area, sulfurous vents represent solfataras and seismic shocks are frequent.

In South Yemen, similar volcanics occur and are predominantly coastal basalt flows and cobble spreads. The most interesting feature at Aden is an extinct crater which is breached on the east and northeast. Its floor is dissected trachytic plateau.

The volcanic series is extremely young as is seen by the state of preservation of many of the ash cones, but extrusion was intermittent over a considerable period back to the Pliocene or even perhaps the Miocene.

To the northwest and east of the Arabian Shield, there was Late Paleozoic rifting which gave rise to a small continent. This comprised the Menderes and Kirsehir Massifs of Turkey and also the Rezaiyeh-Esfandagheh area of Iran and was separated from Eurasia by a northern branch of the Tethys Ocean and from Africa by a southern extension of this. Central Iran may have been a part of Gondwana somewhere between Somalia and India.

The Zagros trough extended from Turkey to the Persian Gulf and is believed to have existed at least from the onset of the Paleozoic until the Alpine Orogeny, the Zagros Mountains being the last spasm of a continuing mountain building cycle. They comprise a suture zone lying between the Arabian Shield and the Persian Plate. This can be divided into five segments. Each of these reflects various lithofacies and tectonic units associated with stages in the collision of these tectonic units. A trench resulted from subduction of the Arabian Shield beneath the stable Persian Plate and endured until the late Middle Cretaceous. It was the site of radiolarian chert deposition, turbidites and olistoliths. There were sharp facies changes in the Late Cretaceous sediments southwest of the deep sea trench. In the northern crush zone,there was rupture and upward thrusting of slices of oceanic crust now manifest as an ophiolitic belt. Later Tertiary movement of the Arabian Shield caused thrust faulting and overfolding next to the trench zone and gentler folding in the belt of simple folds to the southwest.

Also, the Zagros Mountains are gently folded and contain a sequence of Precambrian to Pliocene shelf sediments about 12 km thick which have been folded from Miocene to Recent times. The Cambrian to Middle Tertiary rocks are approximately 7 km in thickness and form a single structural lithic unit called the Competent Group. This is bounded above and below by evaporites and folding due to salt movement may have taken place in the Group itself. This could provide a mechanism for salt diapirism through competent strata and explain how space was made for diapirs, also why they rarely contain country rock relics. The basement did not participate in the folding, but has been deformed by strike-slip faulting (33).

Subsurface faults of Late Eocene to Miocene age exist beneath the Wadi Al-Batin of which the southwest part in Kuwait is heavily dissected, showing evidence of rejuvenation. A regional or local uplift has reactivated the erosional effect of sporadic torrents which had produced the dissected pediment. The Zagros Mountains are structurally linked with those in Oman to comprise an orogen. This includes the Semail Ophiolite Belt of the Oman Mountains which resembles that underlying the Zagros thrust zone of Iran. It suggests that during much of the Mesozoic, an oceanic basin existed. The later destruction of this feature resulted in the emplacement of these scattered remnants of oceanic lithosphere on its former southwestern margin. The process was accompanied by subduction and the tectonic emplacement of the related ophiolites. Harzburgite and dunite from the Oman ophiolite show microstructures reflecting high temperature mantle deformation. A palinspastic reconstruction of the Oman sequence demonstrates a Triassic continent with a marginal ocean basin which underwent subduction on its northeastern periphery during most of the Cretaceous (34).

In the Late Cretaceous, the northern passive margin of the African continent collided with an island arc. Ophiolite obduction was accomplished by continental underthrusting of the fore-arc limb to great depth under an island-arc. The uplift of the ophiolite complex was caused by a combination of compressional shortening and isostatic interaction between the continental crust and the mantle rock which it underthrust.

REFERENCES

1. ROTSTEIN, Y., 1985. Tectonics of the Aegean block: rotation, side arc collision and crustal extension. Tectonophysics, 117, 117-137.

2. ROTSTEIN, Y., BEN-AVRAHAM, Z, 1985. Accretionary processes at subduction zones in the Eastern Mediterranean. Tectonophysics, 112, 551-561.

3. NEEV, D., HALL, J.K., 1984. Mantle produced counterclockwise vortices along the northern Mediterranean belt (a genetic hypothethis for the Alpine System). Mitt. Geol.-Palaeont. Inst. Univ. Hamburg, 56, 111-127.

4. Van HOUTEN, F.B., 1983. Sirte Basin, north-central Libya: Cretaceous rifting above a fixed mantle hotspot? Geology, 11, 115- 118.

5. BROWNE, S.E., FAIRHEAD, J.D. and MOHAMED, I.I., 1985. Gravity study of the White Nile rift, Sudan, and its regional tectonic setting. Tectonophysics, 113, 123-137.

6. BROWNE, S.E., FAIRHEAD, J.D., 1983. Gravity study of the Central African Rift system: A model of continental disruption. 1. The Ngaoundere and Abu Gabra rifts. Tectonophysics, 94, 187-203.

7. BERMINGHAM, P.M., FAIRHEAD, J.D. and STUART, J.W., 1983, Gravity study of the Central African Rift System: A model of continental disruption. 2. The Darfur domal uplift and associated Cenozoic volcanism. Tectonophysics, 94, 205-222.

8. BASAHEL, A.N., BAHAFZALLAH, A., JUX, U. and OMARA, S., 1982. Age and structural setting of a Proto-Red Sea embayment. N. Jb. Geol. Palaeont. Mh., 456-468.

9. LABRECQUE, J.L., ZITELLINI, N., 1985. Continuous sea-floor spreading in Red Sea: An alternative interpretation of magnetic anomaly pattern. AAPG, Bull., 69, 513-524.

10. RESSETAR, R., NAIRN, A.E.M., 1980. Two phases of Cretaceous- Tertiary magmatism in the Egyptian Eastern Desert: Paleomagnetic, and K-Ar evidence. Ann. Geol. Surv. Egypt, 10, 997-1011.

11. BOSELLINI, A., 1986. East Africa continental margins. Geology, 14, 76-78.

12. TAMSETT, D., 1984. Comments on the development of rifts and transform faults during continental breakup; examples from the Gulf of Aden and northern Red Sea. Tectonophysics, 104, 35-46.

13. ABDEL-MONEM, A.A., HEIKEL, M.A., 1981. Major element composition, magma type and tectonic environment of the Mesozoic to Recent basalts, Egypt: A review. Bull. Fac. Earth Sci., Univ. Jeddah, 4, 121-148.

14. BOULOS, F.K., MORGAN, P., HENNIN, S.F., El-SAYED, A.A. and MELIK, Y.S., 1980. The tectonic structure of northeast Africa from gravity data. Ann. Geol. Surv. Egypt, 10, 961-970.

15. BAYOUMI, A.I., BOCTOR, J.G., 1980. A contribution to gravity anomalies in the Gulf of Suez region, Egypt. Ann. Geol. Surv. Egypt, 10, 1027-1035.

16. DAGGETT, P.H., MORGAN, P., BOULOS, F.K., HENNIN, S.F., EL-SHERIF, A.A. and MELIK, Y.S., 1980. Microearthquake studies of the northeastern margin of the African plate. Ann. Geol. Surv. Egypt, 10, 989-996.

17. DAGGETT, P.H., MORGAN, P., BOULOS, F.K., HENNIN, S.F., El-SHERIF, A.A., El-SAYED, A.A., BASTA, N.Z. and MELIK, Y.S., 1986. Seismicity and active tectonics of the Egyptian Red Sea margin and the nothern Red Sea. Tectonophysics, 125, 313-324.

18. BAYER, H.-J., 1985. Kollisions- und Kompressionstektonik am Ostsaum des Golfes von Aqaba. Geol. Rdsch., 74, 599-610.

19. SCHNEIDER, W., PROBST, U. and STÄNICKE, J., 1983. Sedimentological patterns around hydrothermal centers in the Red Sea and Gulf of Aden. Z. deutsch. geol. Ges., 134, 61-93.

20. SEIBOLD, E., BERGER, W.H., 1982. The Sea Floor. 288 pp., Springer-Verlag, Berlin.

21. MORGAN, P., SWANBERG, C.A., BOULOS, F.K., HENNIN, S.F., EL- SAYED, A.A. and BASTA, N.Z., 1980. Geothermal studies in northeast Africa. Ann. Geol. Surv. Egypt, 10, 971-987.

22. HUSSEIN, H.A., El KASSAS, I.A., 1980. Uranium mineralized fault zone at Wadi Um Gir locality, Central Eastern Desert, Egypt. Geol. Rdsch., 69, 567-580.

23. COCHRAN, J.R., 1983. A model for development of Red Sea. AAPG, Bull., 67, 41-69.

24. EPSTEIN, S.A., FRIEDMAN, G.M., 1983. Depositional and diagenetic relationships between Gulf of Elat (Aqaba) and Mesozoic of United States

East Coast Offshore. AAPG, Bull., 67, 953-962.

25. KOVACH, R.L., BEN AVRAHAM, Z., 1985. Gravity anomalies across the Dead Sea rift and comparison with other rift zones. Tectonophysics, 111, 155-162.

26. COURTILLOT, V.E., 1980. Opening of the Gulf of Aden and Afar by progressive tearing. Physics of the Earth and Planet. Int., 22, 343-350.

27. WONG, H.K., DEGENS, E.T.,1984. The crust beneath the Red Sea - Gulf of Aden-East African Rift System: a review. Mitt. Geol.- Pal. Inst. Univ. Hamburg, 56, 53-94.

28. BOUCARUT, M., CLIN, M., POUCHON, P. and THIBAULT, C.,1985. Impact des évènement tectono-volcanique plio- pléistocènes sur la sédimentation en République de Djibouti (Afar Central). Geol. Rdsch., 71, 123-137.

29. MILKEREIT, B., FLÜH, E.R., 1985. Saudi Arabian refraction profile: crustal structure of the Red Sea-Arabian Shield transition. Tectonophysics, 111, 283-298.

30. MOONEY,W.D., GETTINGS, M.E., BLANK, H.R. and HEALY, J. H., 1985. Saudi Arabian seismic-refraction profile: a traveltime interpretation of crustal and upper mantle structure. Tectonophysics, 111, 173-246.

31. PRODEHL, C., 1985. Interpretation of a seismic refraction survey across the Arabian Shield in western Saudi Arabia. Tectonophysics, 111, 247-282.

32. BOHANNON, R.G., 1986. How much divergence has occurred between Africa and Arabia as a result of the opening of the Red Sea? Geology, 14, 510-513.

33. Al-SARAWI, A.M., 1980. Tertiary faulting beneath Wadi Al- Batin (Kuwait). GSA, Bull., 91, 610-618.

34. CHRISTIANSEN, F.G., 1985. Deformation fabric and microstructures in ophiolitic chromitites and host ultramafics, Oman. Geol. Rdsch., 74, 61-76.

Chapter IV-10

DRAINAGE OF A CONTINENT

"Beside the eternal Nile, the Pyramids have risen. Nile shall pursue his changeless way; those Pyramids shall fall; Yea! Not a stone shall stand to tell the spot whereon they stood".
Percy Bysshe Shelley (1792-1822), Queen Mab.

HISTORIC NILE

Hapi was the God of the Nile and its floods, providing water in the desert oases and identified with Nun. He emerged in two whirlpools in the caverns of Elephantine in Egypt after flowing through the heavens and the underworld. His home was on the island of Bigeh at the First Cataract of the Nile. Khnum, the Lord of the Cool Water, was also a deity here and his daughter Satis was the Goddess of the Animating Inundation as well as of Love and Fertility.
One of several ancient Nilometers where the Pharaohs recorded low and high stages of the Nile is at Elephantine. The flood maxima occur late every summer, multiplying the discharge by a factor of nine to reach a volume of just under 9,000 m^3/sec. This is important in connection with estimating agricultural production. In Pharaonic times, this facilitated the assessment of taxes. Then, when the river was rising, navigation on it was forbidden to the King or his Prefects.
The first damming of the river was carried out at the turn of the century by the British. A barrage of red granite rising to 33 m at its crest was sited on the famous multicolored granite basement of the First Cataract, also the source of the many obelisks plundered since antiquity and scattered through Europe. It impounded a reservoir, El Khazzam, extending southward for 200 km. Between 1907 and 1912,

5 m were added to the height and, between 1929 and 1934, a further 10 m were added. This produced a storage of 5 billion m^3 of water, regulated by 180 sluices, extending 350 km upriver to Wadi Halfa on the Sudanese border.

Following the construction of the Aswan High Dam, the Sadd El- Ali, 6 km north of El Khazzam and again on the granites of the First Cataract and the consquent flooding of Lake Nasser, seismic effects resulted and correlate with seasonal fluctuations of the level of water in the reservoir. A recent one was that of November 14 1981 which took place in the Kalabsha area to the west. Resulting from this, a radio-telemetry network surrounding the northern part of the lake was installed and operated since July 1 1982 and supplements three strong-motion stations on the High Dam body. Between July and December 1982, 2,000 events were recorded and their magnitude ranges between 1 and 4.7. The spatial distribution indicates that the activity is confined to three planar zones.

The first is the most active and extends for 14 km along the old Kalabsha fault with foci concentrated in two depth ranges from 14 to 22 km and 4 to 7 km. The second extends for 8 km along another segment of this same fault nearer to the old river bed of the Nile and here the activity is restricted to depths from 4 to 6 km. Hence, a gradual decrease in depth from west to east is apparent. The third is in Wadi Kurkur in the northwestern part of the lake, 15 km south of the High Dam, a zone characterized by poor seismicity. Composite fault-plane solutions showed strike-slip motion for all three zones (1).

This seismic activity is related to the enormous burden imposed by the new dam, which is 4.5 km long and has a crest height of 72 m. Lake Nasser, its impounded reservoir, extends 500 km southward drowning the 12 km long and 2.5 km wide granitic Stone Bed of the Second Cataract under 10 m of water so that ships can pass directly from Aswan to Sudan. The capacity of the reservoir is 130 billion m^3 and there is an estimated 10 billion m^3 annual evaporation over the lake. Its inundation entailed the relocating of more than a score of Pharaonic monuments, including the Ptolemaic Philae. This Temple to Isis is now situated on one of many granitic islands in the Nile between the two dams. The rocks are heavily scoured and fluted and their original colors are masked by a black varnish. Associated whirlpools on the river bed 85 m above sea-level have been abraded to depths of 125 m. The holes so produced are filled with fluviatile sediments. No doubt Herodotus, the 5th Century B.C. "Father of History" was right to accept the story that the local people were deafened by the roar of the waters and the rasping bedload of granite.

THE NILE DELTA AND ITS CONE

Herodotus both recognized and explained the geological phenomena associated with the Nile Delta as well as the lower valley of that river. Since the former resembles the 4th capital letter of the Greek alphabet, he applied its name to it. The priests of Heliopolis told him that all Egypt, except Thebes, was a swamp in the time of the first recorded Pharaoh, Menes, who lived around 3,100 B.C. This convinced him that the Nile Delta and much of the rest of the country was originally covered with lakes and swamps. The areas visited by the Greeks were newly reclaimed land.

The Nile Delta was originally part of the Mediterranean which then extended southward to Asyut. Fluviatile sediments were transported into the sea during its accumulation and, according to Herodotus, lowering of a plumb even one day off-shore produced river muds. It is appropriate that the Herodotus Trough off Cyrenaica and the huge Herodotus Abyssal Plain off the Libyan Plateau are named after this august historian.

The tremendous thickness of the sediments in the Nile Delta are apparent from borehole measurements. In the northwestern delta area, 3,000 m of Quaternary and Pliocene sands and shales were found to be unconformably underlain by more than 1,000 m of Middle Miocene sandy shales and shales. At Lake Manzala, in the northeastern part of the delta not far from Damietta, drilling was carried out to a total depth of more than 4 km. The lowest 1,858 m of this comprised Early to Late Miocene sediments including sands, shales with occasional silts and intercalations of anhydrite. It is zoned using planktonic foraminifera such as Globigerinoides, Globorotalia, Orbulina and Praeorbulina species. Only in the northern part of the Nile Delta is the Late Miocene represented. The upper 2,284 m in the well is occupied mostly by Pliocene shales, silt-stones and sandstones with occasional gravel beds and a capping of Pleistone gravels around 50 m thick.

Drainage of a continent

The Miocene Epoch is characterized by several marine fluctuations. During its early and middle parts, a thick sequence of shale and clay was deposited under the open sea conditions of a marine transgression and contains an appropriate fauna. The Gulf of Suez and the northern Red Sea were invaded from the Mediterranean without any Indo-Pacific subvention. During the Middle Miocene, the Mediterranean connection broadened and the Gulf probably linked with an embayment. Later, shallowing set in and coarse clastics began to predominate in association with a brackish water fauna. This regression resulted from the regional uplift of the Arabian-Nubian Shield, but there may have been periodic communications between the gulf and delta embayments. By the latest Middle Miocene, another marine submergence took place and by the Messinian, marine clays with layers of anhydrite were deposited to form the Rosetta Formation. Subsequently, the northern delta embayment dried up simultaneously with the desiccation of the Mediterranean into a series of terminal saline lakes. This is confirmed by the discovery of stromatolites at 3,460 m depth in at least one borehole. Such cyanophytic algal crusts indicate shallow water conditions of deposition in which photosynthesis must have occurred. In the aftermath of the crisis, a momentous Pliocene event occurred. This was the breaching of the Gibraltar rampart by the Atlantic Ocean to flood into and refill the Mediterranean Basin. Thus, the sea tightened its grip on Egypt, drowning the titanic canyon of the Eonile.

There are several stages in the formation of the Nile Delta. During the Mio-Pliocene, erosion predominated with increasing rainfall. Sediments formed during the Pliocene Epoch are higher than the existing sea level by from 180 to 200 m. Initially, the sea flooded the entire delta which had its southern shoreline confined by a fault system. Sands, silts and shales were deposited to reach a thickness varying from 400 to 1,000 m. The Middle Pliocene beds range from neritic open marine in the north to fluviatile in the south and attain thicknesses of between 300 and 900 m. In the last part of the Epoch, sediments were deposited under the influence of a retreating sea which became isolated from the Gulf of Suez. In the south, the delta dried up and underwent erosion. There was uplift here, in the Gulf of Suez and in the northern part of the Western Desert. The Nile Delta began to form as a result of the increased activity of the river, this process continuing into the Pleistocene. The subsurface part of the Pleistogene succession in the delta reaches a total thickness of nearly 3 km. This has been established west of the Rosetta Distributary.

The actual town of Rosetta became famous in 1799, when a black basalt tablet was found there bearing parallel inscriptions in Greek, demotic and hieroglyphic Egyptian by means of which the last two were interpreted. An important well is at Abu-kir, site of the largest natural gas field in Egypt and another historic location, where Admiral Nelson in 1798 defeated the French fleet supplying the Napoleonic forces in this country. In the north-western delta area, this and other wells show that the Pleistocene is represented by 450 m of sandy shales with occasional gravels and an extremely attenuated covering of Holocene sands.

Sporomorphs found in the Plio-Pleistocene sediments of this offshore well can be grouped into tropical and temperate types. The twenty genera found revealed four paleoecological assemblages. These are grasses and sedges indicating a humid phase, grasses and composite flowers showing a cool dry phase, composite flowers and sedges demonstrating a warm humid phase, sedges and members of the goose-foot family reflecting a dry phase. These may be related to Pleistocene drought and pluvial stages, perhaps also to four eustatic cycles. The exceptional quantity of water handled by the dynamic Nile has converted the delta from a Niger-type to a Mississippi-type so that the deltaic coastline has evolved from an arcuate smooth one into a bird-foot form.

The sea receded in the Plio-Pleistocene and the delta was affected by erosion. Sand, different from that in the Pliocene, was accumulated containing feldspars and gravels of igneous origin. It testifies that the delta contained for the first time sediments carried by waters running from the Red Sea Mountains. The marine regression from the Pliocene estuary started a fluvial phase in which the river cut through sediments filling the bay. This created a terrace and means that Pliocene deposits had completely filled the mouth of the estuary and then had Plio-Pleistocene sand laid over them. Both types of sediment appear on the margins of the present delta, the first at an elevation of 180 to 200 m and the second as a terrace between 90 and 120 m above the height of the delta (2).

During the Pleistocene, there were several eustatic fluctuations of the sea-level so that sometimes the delta was flooded. Pleistocene formations are also found as terraces on both sides of it at elevations between 30 and 45 m. Deposits of the Early Pleistocene show that the Nile at that time started to meander and be deflected westward, leaving areas of sediment on the east bank which were abandoned when the river started deepening its course. The Middle Pleistocene deposits are silts and typical products of aggradation. When sedimentation ceased, the river resumed deepening its course through the delta. At that time, the Mediterranean was 12 m lower than now. Subsequently, sea-level rose with accompanying aggradation in the delta, this rise proceeding until it stabilized in historic times at the existing level. The textural characteristics of the coastal sands show that the central part of the coast in this area has advanced, marine sediments up to 22 m in thickness having been lain down over the backshore flood-plain. The distribution of translucent heavy minerals is controlled by hydrodynamic forces. There are four characteristic suites which may be useful in differentiating between breaker zone, beach, backshore and dunes. The continental margin of Israel was formed by accumulation of fine clastics coming from the Nile by northeastern currents of the Mediterranean since the Pliocene. After deposition, the material was redistributed by earthquake-induced slumping and sliding (3,4).

Geophysical investigations at sea off the delta show recent sedimentation and tectonism, especially the effects of salt deformation resulting from movement of a Late Miocene sequence. The continental rise can be divided into the Nile Cone and the Levant Platform. The second of these comprises an irregular seafloor morphology extending north from the Sinai. It separates the cone on the west from the Cyprus Basin on the east. Curiously, postulated off-shore fans of the Rosetta and Damietta distributaries of the present mouth of the Nile do not exist. Rather, there is only one major feature - the Nile Cone.

This is one of the major submarine delta fans of the world at 320 km wide and extending out to a depth of 2,800 m on the west and 1,600 m on the east. Drilling at a DSDP site located at the extremity of the western Nile Cone showed 270 m of interbedded Quaternary sands and clays. These imply that the cone was rapidly built up from Nile sediments at a rate of 0.5 mm annually. Calculations of thicknesses are based on the assumption of an average seismic velocity of 2 km/sec in the sediment. The cone has such a vast volume that its accumulation must have commenced in the Miocene if existing erosion rates prevailed in the past. Allowing for crustal sag in late Cenozoic times, the total volume of the cone and delta is believed to be some 220,000 km³. There is a concordance between the volume of the Nile Cone and the estimated volume of rock removed from Ethiopia suggesting that the Nile Cone derived its material from there. The modern Nile erosion values are very high, perhaps 120 million tons per year for the main Nile, reflecting a doubling in annual load during the 70 years before the completion of the Aswan High Dam (5).

Both this and the Levant Platform have great thicknesses of Nilotic sediment. The irregular sea floor of the latter is the result of greater vertical and horizontal flow of evaporites than occurs on the Nile Cone. In fact, evaporite movements caused a considerable number of collapse structures as well as a 100 km-long salt ridge at the northern termination of the platform.

One of the seismic profile reflectors is either a Middle to Late Miocene carbonate sequence which prograded eastwards during the Messinian regression or an erosional surface cut into pre-Messinian strata. This reflector denotes the top of an elongate anticlinal structure off the Nile Delta and Sinai. The volume of post-Messinian, Nile-derived sediment is almost 400,000 km³, i.e. an average sediment thickness of 1.9 km for the area and a mean sedimentation rate of 3.7 cm per century.

As for plate tectonics in this region, it is quite unneccessary to infer a Levantine Plate between the present Arabian and Eurasian ones. It has been suggested that the "Pelusium Line" may be the western boundary of a proposed "Central Plate" wedged between surrounding plates and causing compressional and complementary tensional features in much of the Middle East. However, there is very little evidence for such a plate.

THE UPPER NILE

At the end of the Eocene Epoch about 40 Ma, the Tethys Ocean regressed. Oligocene rivers began their erosional activities. Thus, the oldest Tertiary drainage system began in response to tremendous tectonic movements which elevated the Arabian-Nubian Shield and initiated sea-floor spreading in the Red Sea. The shorelines of the latest Eocene from the Fayum to Siwa Oases are bedecked with the scattered

Figure 10.1: Silicified driftwood in fluviatile gravels of Oligocene age in Wadi Natrun, Western Desert.

remains of fossil vertebrates. Clastics began to intrude upon the previously predominant carbonate sedimentation and this process continued more vigorously in the Oligocene. Sands, cobbles and petrified wood can be identified from this prolonged humid episode (Figure 10.1). The Epoch was one of erosion in Egypt so that a prominent unconformity resulted. Its drainage system developed over a relief quite different from that of modern Egypt and one partly preserved as gravel ridges. There was a major river, the Urnile, which entered the Mediterranean at the eastern tip of the Qattara Depression. Between 38 Ma to 5 Ma ago, i.e. from the Oligocene through the Miocene, there was extensive volcanism and tectonic disturbance. The cutting of the Kurkur Pediplane at 300 to 360 m occurred in Lower Nubia under semi-arid conditions in the Early Miocene.

By far the most spectacular phenomenon of the Late Miocene was the isolation and desiccation of the Mediterranean Sea, the renowned salinity crisis, accompanied by the intense erosion of Egypt and the excavation of the Nile Valley. Much of the Late Miocene river system seems to be deeply buried, e.g.

along the Mamarica Plateau in Libya, and its deposits accumulated over the exhumed surface of an erosional episode of this age. These very deposits may have been bevelled by later erosion to create sand sheets.

Figure 10.2: Distribution of marine Neogene in Egypt showing the great extension of the Eonile estuary.

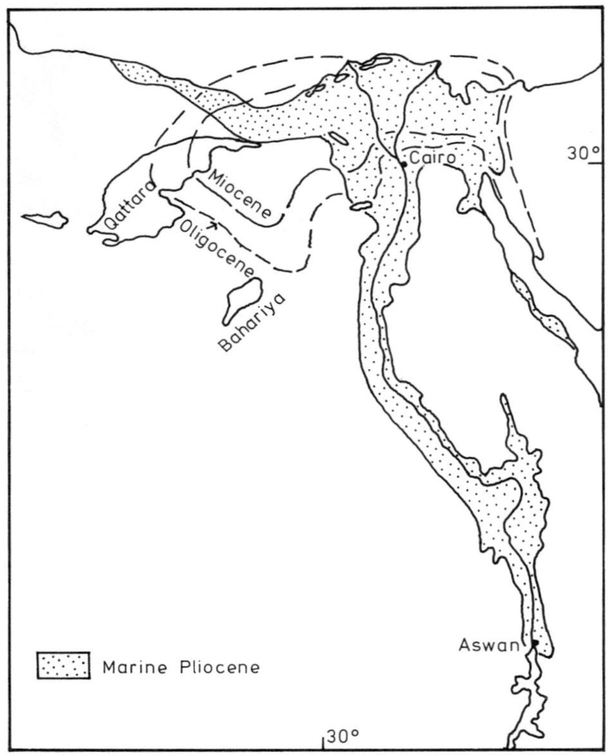

In the Late Miocene when the Mediterranean dried up, a river termed the Eonile began to form. Since there was a considerable drop in base level, it cut an impressive canyon from the vicinity of the existing Nile Delta to Aswan. Its dimensions compared with those of the 320 km long Grand Canyon of the Colorado River, but its shape was quite different. It was four times longer and even deeper. The Eonile Canyon reached a maximum depth of 2,500 m near Cairo compared with 2,080 m for the Grand Canyon. Entrenchment into bedrock near Cairo was 568 m below modern sea-level and near Aswan, it reached 172 m below this datum (Figure 10.2). Both the Colorado and the modern Nile traverse large tracts of desert, but obtain enough water from sources beyond to maintain their supplies. They carry roughly equal loads of sediment, 300,000 tons daily. However, the Eonile was a much more dynamic river which deposited perhaps 70,000 km³ of sediments in the space of less than 3 Ma, i.e. 20% of the total mass of material in the Nile Cone. The gradient of the Eonile averaged 1:400 from Aswan to Cairo and 1:250 in the Asyut to Cairo section. For comparison, the Nile today has a gradient of 1:13,000 from Aswan to Cairo (6,7).

Both the Colorado and the Nile are antecedent rivers, but the reasons for this are different. The Colorado results from rejuvenation of the landscape, whereas the Nile is caused by continuous lowering of the base level. This reflects the differences of the continental margins, that of the Nile relating to a subducting and northward-moving African Plate. Furthermore, the Nile was cut rapidly during the Late Miocene and is therefore appreciably younger than the Colorado which formed from the Early Miocene until the present.

Figure 10.3: A reconstruction of Anancus osiris.

The Early Pliocene, about 5 to 3 Ma ago, was characterized by a refilling of the Mediterranean Sea so that the Eonile Canyon was transformed into a long estuary which became filled up with marine sediments to approximately one third of its depth as far south as Aswan. The Late Pliocene, 3 to 2 Ma ago, had a cool wet climate in which the Paleonile formed and converted the estuary into a channel. The previous canyon ultimately became filled up through aggradation by the Paleonile system and one-fifth of the fluvial sediments in the Nile Valley were deposited by this Pliocene river. The favorable climatic conditions of abundant moisture facilitated the colonization of the area by the Pliocene mastodon, Anancus osiris (Figure 10.3), recovered from Gizeh (8).

Egypt became an increasingly arid desert in the Early Pleistocene and the Paleonile stopped flowing. Towards the end of this time interval, perhaps as much as 1,800,000 years ago, a rather short-lived, highly energetic river, possibly stimulated by the Idfuan Pluvial, the Protonile formed. This deposited distinctive cobbles and gravels in a course in the Nile Valley, but 10 to 15 km to the west and sometimes as much as 100 m above the modern floodplain. The sources of this river are not known, but probably were the same as those of the Paleonile.

From the Armantian Pluvial, 1.4 to 1.2 Ma ago, the earliest pebble tools so far known from the lower Nile Valley derive and come from the cliffs of Abu Simbel. This is across the river from the world-famous temple of Rameses II, now relocated at a higher elevation on the adjacent plateau,and also from the Armant Formation gravels near Luxor. Massive travertines deposited during this pluvial are overlain by breccias, commercially called brocatelli, which have been indurated by red soil to form an ornamental rock exploited since antiquity.

The break-through of the later Prenile into Egypt resulted from stream captures in the Ethiopian Highlands. This highly energetic river excavated a new route separate from that of the older Niles but still within the confines of the present Nile Valley. Confirmation is provided by the deposition of clastics containing epidote and pyroxene coming from streams flowing from east Central African sources. This is the first appearance of these minerals in fluviatile deposits in Lower Egypt. Relative percentages of the 2 minerals in Prenile sediments are only slightly different from those in modern Nile deposits.

The Nile drainage basin enlarged through the successive capture of Wadi Atbara and the Blue Nile by the Prenile. A period of lesser aridity marked by gravels containing Achéulian tools which profoundly affected the Arbain Desert to the west was accompanied by the cessation of the Prenile. Intense rains characterized the succeeding Abbassian Pluvial and laid down huge accumulations of gravel deposits, the first generation of gravels which came from the uncovered basement rocks of the Eastern Desert of Egypt. They comprise probably the most conspicuous horizon in the Nile sequence and are rich in archeological material and Late Acheulian artifacts. This culture ended 300,000 years ago. One of its sites, Arkin 8 (near Wadi Halfa), yielded nearly 3,000 artifacts from only one-eighth of its area and possesses the earliest known shelters in Egypt and the Sudan. These are the earliest dwellings in the world next to the basalt block ring of the Leakeys at Olduvai, Tanzania.

Lower Paleolithic sites along the main Nile from Khartoum to the delta clearly show habitation soon after the emergence of species of Man in East Africa and Ethiopia. The distribution of sites in the Wadi Halfa area, only a few kilometers from the river, indicates that Early to Middle Acheulian peoples used the resources of the river. Another important site is BS-14 situated in the Libyan Desert at Bir Sahara Depression which supported artesian lakes during the Late Acheulian and Mousterian from roughly 300,000 to 100,000 years ago.

Figure 10.4: An Acheulian handaxe exposed by deflation on the desert floor of the Great Sand Sea.

Many eggshells and bones have been found which imply that Late Acheulian people migrated on to the Saharan savanna during the Abbassian Pluvial. It prevailed for almost 100,000 years between 300,000 and 200,000 years ago, succeeding hundreds of thousands of years of drought. The restricted pasturage and few water holes together with the seasonal rainfall available may have established conditions forcing Man, who was then making a transition between Homo erectus and Homo sapiens neanderthalensis, to travel seasonally on the open grasslands and retreat to permanent oases during the dry seasons. Around such springs, fossil pollen and plant remains indicate that plants were plentiful at oases such as Kharga during the Paleolithic pluvials (Figure 10.4).

The final Acheulian people enjoyed the Saharan I Pluvial while the deposits of the succeeding Saharan II Pluvial contain Mousterian artifacts. It must be mentioned that the Saharan III Pluvial has Middle Paleolithic Aterian implements associated with it, these being probably 80,000 or so years old. This culture coexisted with the aggradation of the mixed Nilotic-wadi deposits of the Korosko Formation. These grade upstream into alluvial terraces perhaps related to widespread colluvial wash on pediments next to the wadi floors. The Aterian is not found in the valley of the Nile, apparently being desert orientated. It is found spread through the entire Sahara north of the 15° N parallel extending from the Atlantic coast in the west to the Western Desert oases. However, it never penetrated Cyrenaica and is rare in Tripolitania (9).

A different scale record of Pleistocene paleoclimates from those in the Nile Valley fluviatile units is afforded by a sequence of alluvial and spring-fed subaqueous deposits above and below the great Libyan limestone cuesta in the upper reaches of the Wadi Kurkur, 60 km southwest of Aswan. After the Pliocene Plateau Tufa, four complexes of similar calcareous materials formed in the Oasis and as fans below the escarpment. They range in age from more than 40,000 to less than 32,000 years, the youngest possessing unworn artifacts which record a major Aterian occupation. This was contemporary with the later member of the Korosko Formation, e.g. on the Kom Ombo Plain. From the Kurkur sequence, it may be inferred that the Abbassian Pluvial was much more effective, longer and wetter than its successors in the terminal Pleistocene and Holocene, the period of intense rainfall alluded to above.

The successor of the Prenile was the weak Neonile, with four stages, of which the early history is distinguished by a long interval of recession and minor flow. This latter can be attributed to the depletion of water from the hydrologic cycle as polar ice during the contemporary Würm Glaciation. This was the last and biggest in the Pleistocene and dropped the sea-level at least 120 m. Between 20,000 and 17,000 B.P., the river floods were higher than recent ones and there was accelerated water action in wadis, the hyperaridity dominant since the end of the Mousterian was being moderated. This phase is termed the Masmas Formation, representing a continuous sequence of periodic Nile aggradations over 50 m thick. It was accompanied by a higher groundwater table which created seasonal mini-oases. There is reason to believe that this may be connected with the Kubbaniya Pluvial named after silts containing Middle Paleolithic artifacts in the Wadi Kubbaniya west of Aswan. Simultaneously, under semi-arid conditions,the Ballana Pediplane was cut at 230 to 260 m in Lower Nubia.

Subsequently, a recessional period took place and the floods abated. There followed a new phase of high seasonal floods. The Sahaba-Darau Aggradation commenced, attaining a maximum around 15,000 B.P. and lasting until 12,000 B.P. However, the climate was basically arid with brief periods of rainfall and there were several major stone industries. One of these is the Qadan in Lower Nubia about 250 km upriver from Aswan of which cemeteries have been found. Maybe children under three were excluded, since none have been found. The interred adults showed a variety of pathologies including arthritis, osteitis, possible spinal tuberculosis, sacro-iliac problems and many dental disorders such as missing teeth, cavities and abscesses. Nearly half the burials possess small flake points, some nothing more than burins, formerly points of arrow shafts. Extensive violence may be inferred and no doubt reflected the fight for limited resources and living space. Since both sexes are represented among the skeletons, the violent deaths most likely resulted from ambushes rather than organized warefare among young adult men.

Many geologists and archeologists believed that Paleolithic material would not be found on the flood-plain of the Nile as a result of burial or washing away during the years. Nevertheless, in 1968, a well-stratified Paleolithic site was spotted at El Kab on the eastern bank of the Nile hemmed in by the Red Sea Hills. Microlithic tools were detected and are of Epipaleolithic age. This period of small-blade technology dates from the 9th and 10th millenia B.P. and lacks ceramics. Indeed, very few of its sites are stratified or dated and they are much less frequent than Aterian and Neolithic sites. Probably the human population at the time was generally rather thinly distributed in comparatively small groups. Excavations in the flood-plain located

Predynastic burials and earlier occupation levels, the oldest radiocarbon-dated to 8,400 B.P. and yielding nearby artifacts. This was the time of the pre-pastoral Neolithic or older Neolithic cultural entities.

Probably owing to its capture of the White Nile approximately 10,000 years ago, the Neonile came to resemble the modern river. This seminal event produced practically total integration of what is now the longest drainage system in the world extending from the Mediterranean to Lake Victoria, a distance of 6,695 km with 1,450 km of it in Egypt. Where the Blue Nile, the White Nile and the Wadi Atbara went before they were captured by the Egyptian Nile is unknown. The key to its solution may lie in the paleodrainage of the Arbain Desert as revealed by SIR-A.

Information regarding the geological history of the divide between the Nile and the Chad Basins is very sparse. To the south, a structural warping of the Gebel Marra basement rocks created a dome about 500 m high with a 140 km long superimposed volcanic complex which started erupting in the Miocene and continues with solfataric activity until now. This has promoted major and minor diversions in the regional drainage systems. The western end of Wadi Howar may have been captured so that 60,000 km² of catchment area of the Nile Basin was diverted to the Chad Basin.

Similarities between the fish fauna of the Nile River and Lake Chad imply that drainage from an ancestral Upper Nile connected with that of Lake Chad. If true, such a connection would solve the problem of explaining what happened to the older equatorial headwater tributaries of the existing Nile before they were captured, successively from east to west, during Middle to Late Pleistocene times. An old suggestion that drainage from Lake Chad may have reached the Upper Nile at Dongola through Wadi Howar could be consistent with the known eastward extent of this lake during the Quaternary pluvials when it rose to elevations of at least 320 m above the existing level. Wadi Howar is a dry stream course which extends eastward towards the River Nile across Dafur in northern Sudan, passing some 650 km south of the Gilf Kebir Plateau and about 200 km north of Gebel Marra. It marks the southern boundary of the Arbain Desert.

The modern Nile derives its waters almost entirely outside Egypt and drains more than 3 million km³. In the same way, sections of the Blue Nile and Wadi

Atbara are remarkably recurved and look as if they were forced westward as they found a way around shield volcanoes, volcanic piles and uplifts which were emplaced during the Tertiary. In fact, the Nile too describes a great bend in the middle Sudan around the Bayuda Volcanics which mark the end of the Central African Rift System bearing southwest from them through the Gebel Marra complex to Mount Cameroon.

The Gebel Marra basaltic-trachytic strato-volcano is located at the intersection of this system with a northwesterly-directed line of structures and volcanics passing through the Tibesti complex and the Gebel al Haruj to the Gebel es Soda volcanics and the Nefusa Uplift near Tripoli. Clearly, the evolution of the Gebel Marra and the Tibesti volcanic complex together with the Gebel Uweinat at the "Dreilaendereck" between Egypt, Libya and Sudan must have profoundly influenced the pre-Pleistocene regional drainage of the eastern Sahara.

North of Tibesti lies the Gebel al Haruj, one of the most voluminous basalt provinces in North Africa. It contains volcanics ranging in composition from olivine basalt to hawaiite and mostly porphyritic with phenocrysts of olivine and, to a lesser degree, clinopyroxene. The area involved is 45,000 km² and two post-Oligocene episodes of basaltic volcanism took place. The first produced a lava plateau and the second, believed to be of Late Pleistocene age, formed a set of more restricted volcanic centers which are mostly in the central part of the main lava plateau. Radioisotope dating indicates that Al Haruj volcanicity went on from 6 to 0.4 Ma ago, most of the flows being younger than 2.2 Ma years (10).

About 150 km northwest of Gebel al Haruj is the volcanic complex of Gebel es Soda situated in Central Libya and covering 6,000 km². This comprises a plateau made up of a series of lava flows from which younger volcanic cones and shield volcanoes emerged. Most of the rocks correspond to olivine and alkaline basalt. Radioisotope ages from its oldest basalts range from 12.3 to 10.5 Ma, corresponding with Middle to Late Miocene and partly agreeing with paleomagnetic studies which show that the basalts of the Gebel es Soda are of Miocene to Pliocene age (11).

In northwestern Libya, Tertiary-Quaternary volcanic activity in the Gharyan area consists of three quite distinct episodes. The first comprises plateau lavas, 55 to 50 Ma old, covering an area of 3,000 km² southeast of Gharyan. They have a thickness of a few tens of meters with never more than four or five flows occurring in one place. The lavas are very uniform in composition, close to the basalt-hawaiite boundary, and have characteristics transitional between alkalic and tholeiitic affinities. The second is composed of phonolite domes, 40 Ma of age, lying to the north and northwest of the lava plateau and forming hills above the Mesozoic limestone terrain. They show various different field relationships to the country rock, from laccolithic to extrusive domes and all are strongly undersaturated and peralkaline. The third includes late volcanic centers, 12 to 1 Ma old, representing minor volcanic vents. They form hills or groups of hills on top of the lava plateau. Away from the plateau, the centers form positive, easily recognizable features which, being dark, stand out against the white Mesozoic limestones (12).

Some people have proposed an evolutionary relationship between these episodes like that observed in oceanic island volcanoes such as Hawaii, in which case the late volcanic centers could be considered as a rejuvenation phase. However, caution must be exercised because of the much greater time-scale at Gharyan and the much longer intervals of time between the sharply defined volcanic episodes of the Libyan activity.

The Libyan volcanic rocks constitute a meridional belt which commenced 12 Ma ago in the Middle Miocene when Red Sea spreading was initiated and similar tensional movements are manifested in the Hun Graben between the Gebel es Soda and the Gharyan area which is infilled with lacustrine sediments of probable Oligocene age.

REFERENCES

1. SIMPSON, D.W., KEBEASY, R.M., MAAMOUN, M., IBRAHIM, E.M. and ALBERT, R.N., 1985. Induced seismicity around Aswan Lake. Tectonophysics, 118, 281.

2. EL FISHAWI, N.M., 1985. Textural characteristics of the Nile Delta coastal sands: An application in reconstructing the depositional environments. Acta Min.-Pet., Hungary, 27, 71-88.

3. EL FISHAWI, N.M., MOLNAR, B., 1985. Mineralogical relationships between the Nile Delta coastal sands. Acta Min.-Pet., Hungary, 27, 89-100.

4. ALMAGOR, G., GARFUNKEL, Z., 1979. Submarine slumping in continental margin of Israel and northern Sinai. AAPG, Bull., 63, 324-340.

5. ROSS, D.A., UCHUPI, E., 1977. Structure and sedimentary history of southeastern Mediterranean Sea - Nile Cone Area. AAPG., Bull., 61, 872-902.

6. McCAULEY, J.F., SCHABER, G.G., BREED, C.S., GROLIER, M.G., HAYNES, C.V., ISSAWI, B., ELACHI, C. and BLOM, R., 1982. Subsurface valleys and geoarcheology of the Eastern Sahara revealed by Shuttle Radar. Science, 218, 1004-1020.

7. BENTZ, F.P., HUGHES, J.B., 1981. New reflection seismic evidence of a Late Miocene Nile Canyon. In: The Geological Evolution of the River Nile by SAID, R., 131-138, Springer, New York, USA.

8. SAID, R., 1981. The geological evolution of the River Nile. 151 pp., Springer, New York, USA.

9. BUTZER, K., 1981. Pleistocene history of the Nile Valley in Egypt and Lower Nubia. In: The Sahara and the Nile, ed. WILLIAMS, M.A.J., FAURE, H., 253-280, Balkema, Rotterdam, NL.

10. BUSREWIL, M.T., WADSWORTH, W.J., 1980. Preliminary chemical data on the volcanic rocks of Al Haruj area, central Libya. In The Geology of Libya, 3, ed. SALEM, M.J., BUSREWIL, M.T., 1077-1080, Acad. Press, London, UK.

11. WOLLER, F., FEDIUK, F.,1980. Volcanic rocks of Jabal as Sawda. In: The Geology of Libya, 3, ed. SALEM, M.J., BUSREWIL, M.T., 1081- 1093, Acad. Press, London, UK.

12. BUSREWIL, M.T., WADSWORTH, W.J., 1980. The basanitic volcanoes of the Gharyan area, NW Libya. In: The Geology of Libya, 3, ed. SALEM, M.J., BUSREWIL, M.T., 1095- 1105, Acad. Press, London, UK.

Chapter IV-11

ONSET OF ARIDITY

"The world is a very ancient battered tome, in which its tale is writ: Alas! the first and the concluding pages have fallen out of it".
Mirza Muhammad Ali, "Saib", 1605-1677.

THE MIOCENE CRISIS OF THE MEDITERRANEAN

The northward convergence of Africa on Europe during and after the Mesozoic had many significant effects. In the Jurassic and Cretaceous, the tropical Tethys Ocean divided the two continents. During the Late Cretaceous, Africa joined Europe and a shallow shelf sea linked the Atlantic and Indian Oceans. This disappeared in the course of the later Early Miocene 20 Ma ago, this principal paleogeographic event in the Tertiary creating the existing pattern of land and sea. The Mediterranean remained connected with the eastern European Paratethys inland sea by a constricted seaway north of the Alps until the early Middle Miocene when uplift started. About 15 to 14 Ma ago, during the Late Langhian and Serravallian Ages of the Middle Miocene Epoch, the Mediterranean finally separated from the Paratethys so that no saline water flowed from it into the latter and there was no return flow of freshwater (1).
Through this epoch, evaporites were deposited in a marine cut-off in northern Libya. One 200 km^2 outcrop is 30 km east of Benghazi. It comprises coarsely crystalline gypsum which reaches a thickness of 10 to 12 m and may foreshadow the Messinian desiccation crisis of the Mediterranean. The conditions of deposition must have involved alternate flooding and evaporation in a confined basin leading to saturation of water with respect to gypsum. This is inferred from the occurrence of detrital laminae in optically continuous crystals. Interestingly, in South Australia, similar ones possess delicate laminae resembling these. In fact, such crystals in

artificial salinas there are identical to those of the Libyan Miocene, except in size. Later, fine detrital carbonate material carried in suspension with the flood water must have deposited and afterwards, evaporation produced concentration of brines. Further crystallization of calcium sulfate then took place in optical continuity with the earlier crystals.

Evaporites are restricted to the central and north-eastern facies belts in Libya, gypsum and anhydrite occuring together with clays and carbonates in the upper 150 m in the Marada Formation of the Early and Middle Miocene (2).

Around the time interval 8.5 Ma to 7.5 Ma ago, widespread deposition of evaporites up to 1,000 m thick took place throughout the Mediterranean Basin and this implies that there was a continuous brine supply coupled with intense evaporation. At the end of this time, the Antarctic ice cap expanded to lower the sea-level all over the world by 40 m and cut off the available brine. Consequently, the Mediterranean was converted into a low-lying salt desert resembling the Dallol salt plains of the northern Afar Depression, but on a much larger scale. Thus, there was a final isolation from the Atlantic by the Late Miocene 6.5 Ma ago. No doubt, animals were able to cross this salt desert and any remaining plants must have adapted to the aridity.

Peripheral rivers cut very deep canyons along the Mediterranean shoreline and redeposited some of the original salts. Several hundred meters of younger evaporites, including dolomite, gypsum and anhydrite, were precipitated simultaneously. With such low base levels, the Nile Cone and Delta became transformed into a subaerial alluvial fan at the distal parts of which evaporites accumulated. The River Nile entrenched itself very deeply into bed rock reaching 570 m below present sea-level at Cairo and 172 m below this datum near Aswan. Although epeirogeny took place and fracture zones existed, nevertheless the excavation was mainly a fluvial response to this extraordinary lowering of the base level. It is apparent, therefore, that the Mediterranean Basin was segregated from the world oceans during the Late Miocene and extensive evaporitic deposition clearly demonstrates that there was no Mediterranean Sea between 8 and 5 Ma B.P. (3).

During Early and Middle Miocene times, the ostracods of the rapidly changing western Tethys-Mediterranean Sea contained some genera typical of the margins of the world ocean basins implying that some of the inner Tethyan basins had open access to the deeper world ocean fauna. Continuity was broken in the late Miocene Messinian age when widespread evaporite deposition took place and was followed by localized continental lagoonal deposition.

These late Miocene events involved the catastrophic environmental change of the salinity crisis discussed above which took the form of a profound cooling from 6 to 5 Ma ago and are sometimes referred to as the Terminal Miocene Event. A parallel abrupt climatic change from warm to cold took place. Evidence for this is provided by deep sea-core and terrestrial deposit analyses as well as by fossil pollen, land snails and dramatic alterations in the lives of mammals. The great drop in temperatures built up so much ice in Antarctica that the worldwide sea level dropped about 60 m and the rainfall in many places was strongly influenced. Vast regions of tropical Africa, formerly warm and wet, turned cool and arid. Forests retreated steadily and were displaced by advancing grasslands and clumps of low trees as in the present-day savanas of East Africa. The habitats of the apes diminished alarmingly. At the same time, great associated changes took place in African fauna. On the one hand, some forest-loving antelope species vanished completely while, on the other, grazing antelopes appeared. This means that the harsher environmental conditions stimulated extinction of some groups and an explosive radiation of new species adapted to them. No doubt, among the newcomers, were the earliest australopithecines, the ancestors of the Taung child and possibly modern Man.

The western Antarctic ice sheet partly melted with a resulting rise in sea-level which flowed over the barrier at Gibraltar, the natural dam excluding Atlantic waters from the desiccated Mediterranean Depression. After this collapse of the western threshold which had divided the newly formed basins of the Mediterranean from those of the Atlantic, open ocean waters flooded into the deeper western central basins. Thus, marine conditions returned, terminating the salinity crisis.

This Pliocene invasion brought in ostracods from the deeper (psychospheric) fauna from well below the thermocline as far eastward as Calabria. The changes between these major episodes happened in a probable time span of 2 Ma. Ostracods indicating continental conditions are sometimes immediately succeeded by those showing connections with deeper parts of the

Atlantic. As there are no intermediate shallow marine (thermospheric) faunas, extremely rapid, almost catastrophic, changes must have occurred in the western Mediterranean between these episodes.

The Pliocene sea submerged the evaporites beneath 1,500 m of water and embayment of it encroached into the lower Nile Canyon. Estuarine sediments were deposited in the drowned Nile Valley as far as Aswan and it can be assumed that Egypt was much more humid than it is now. At about the same time, the Indian Ocean linked with the Red Sea across the Bab el Mandeb Straits through glacio-eustatic rising sea-level.

In the overdeepened Nile trench, a 135 m thick suite of montmorillonite clays and interbedded sands with rich organic debris, glauconites and zeolites was lain down and is now preserved at elevations ranging from 172 to 35 m below sea-level as far south as Aswan. Occasional ostracods are found and imply brackish water estuarine conditions. These Early Pliocene sediments at Aswan are followed by over 100 m of sands with gravels and clay lenticles and, near Kom Ombo, by a minimum of 40 m of sands or shales with gypsum intercalations. In the limestone parts of the valley of the Nile between Luxor and Fayum, there are more than 60 m of calcreted quartz sands, marls with gypsum and cherty or limestone conglomerates showing complex facies changes. These are unfossiliferous and best preserved in embayments of the Miocene Nile trench. In the Valley of the Kings, there are conglomeratic and brecciated units of the Issawia Formation at 180 m elevation. This compares with a maximum of 130 m for Pliocene deposits at Cairo or Kom Ombo. The former may be younger than a sand unit at Aswan, but the lowest part of the sequence, the Armant Formation, is a Late Pliocene facies complex coeval with Ostrea-Pecten beds west of Cairo at 121 m elevation. It is interesting that the Gizeh Pliocene contains the elephant Anancus osiris. In sum, these Late Pliocene beds confirm abundant moisture in Egypt at that time and also a degree of torrential run-off.

There are Late Pliocene valley margin tufas which may be represented in the Libyan Tableland by the Plateau Tufas of Kurkur and Kharga. At Kurkur, there are two units exceeding 45 m in thickness and having basal clastics and occasional organic tufas in an extensive deposit of travertine-like, laminated freshwater limestones. The sediments were deposited near an earlier drainage line and they contain impressions of leaves of Ficus together with a wide range of pollen, especially palms and Acacia. This suggests a sub-humid climate with adequate supplies of water.

THE WESTERN DESERT

Did the Mediterranean salinity crisis during the Messinian have any Paleogene precursors? This question may be answered in the affirmative as is evidenced on the western and eastern flanks of the Arabian-Nubian Shield. The Oligocene in the eastern Sirte Basin reaches a maximum thickness of 720 m of both fossiliferous marine and non-fossiliferous continental facies, the latter restricted to the southwest and south. In the former, glauconitic calcareous sandstones, limestones, dolomites and clays with some evaporites are to be found, this sequence comprising the bulk of the sediments of this period. The Oligocene is thickest in the north-western structural trough and thins to the south and southwest. Apparently, differential subsidence exerted an important influence on the sedimentational processes, thicknesses increasing in structural lows.

The Early Eocene of the eastern Sirte Basin consists of 570 m of evaporitic interbeds of dolomite and anhydrite with sporadic limestones, the basal 30 m of which is a fossiliferous limestone depositionally related to the underlying Paleocene. Most of the evaporite sequence is poorly fossiliferous, but deposition must have occurred in a restricted shallow marine environment. The Middle Eocene is represented by a 420 m thick nummulitic limestone succeeded by 75 to 90 m of sandy dolomitic limestone, sandstone and claystone transitional to the overlying Oligocene.

It might be expected that this Middle and Late Paleogene salinity crisis should be reflected on the northern side of the Tethys Ocean. In fact, this is the case. Perhaps the best instance is that of the Upper Rhine Valley in Central Europe where a number of basins developed in the Early Eocene. These became infilled with lacustrine sediments of which the most important used to be the 200 m thick Messel oil-shales, formerly exploited and now abandoned, but still famous because of their vertebrate fossil contents. The lacustrine sedimentation continued and brackish-water deposits of the Pechelbronn Formation were lain down. Oil also exists in this and has been extracted since the 15th Century. In addition, the 1,300 m thick formation contains 150 m of evaporites, including halite and sylvite. As will be seen below, the Miocene salinity crisis of the Mediterranean similarly spread around the peripheral countries over a similar duration. The northern manifestation of the saline conditions in this case is best exemplified by the Carpathian Foreland. Here, in the Serravallian Age, salt deposited from the beginning of the Epoch reached its maximum development.

In Europe and the Levant, the Neolithic involves the appearance of various aspects of material culture such as ceramics and stone-ground tools. On the other hand, in the Sahara this is meaningless because ceramics appear far earlier than the initial indication of domestic plants or animals. Therefore, it is believed that the development of the culture in the latter was a dynamic process lasting at least 3,000 years. Domestic animals first appeared in the Sahara just after 7000 B.P. Despite this, hunting wild animals and collecting wild plants still went on. At this early stage of food production and uncertain weather conditions, it would have been extremely unwise to depend on any one source of food. Unquestionable, at this time, the Western Desert was far from the mammalian Garden of Eden described above with lush vegetation and teeming game. Nevertheless, rather agreeable savanna conditions prevailed with adequate numbers of perennial lakes which have left thick beds of diatomites in areas incapable of supporting animal or plant life today. The earlier human exploitation of these lakes is evidenced by the remains of aquatic animals hunted and fished with bone harpoons as well as pottery. This latter bears a decorative motif, the dotted wavy line, implying a cultural link with similar ware found in other places such as the confluence of the Blue and White Niles and the Central Sahara, e.g. north of Timbuktu. An unusual method was used to produce this ornamentation. A catfish spine was dragged across the unfired clay and could be varied to produce punctate undulating lines as well as comb and even roulette decorations. The region continued to be occupied by pastoral people until 4500 B.P., when droughts became common and they were forced to retreat to permanent or seasonal rivers (4,5).

At the beginning of the Neolithic rainy period, the time of the terminal Paleolithic-Neolithic Pluvial, the Dishnian, which lasted from 10,000 B.P. to 5,500

B.P., probably 25 cm of rain fell annually. This must have enriched the surrounding environment so that plant and animal life flourished in abundance. The amount is very small, but still more than Cairo receives today (less than 5 cm per annum). Even at this figure, twenty-two separate plant species were recorded in 1952 at the Wadi Gebel Ahmar (the Red Mountain Gulch). There are no tamarisk groves at this site probably because of its proximity to the urban center of Cairo, the citizens of which have destroyed them for cheap fuel and pasturage. It is now known that timing was a crucial factor in the development of Predynastic Upper Egyptian culture which was balanced on a precarious acme of climatic optimum. Between 9,000 B.P. (7000 B.C.) and 2500 B.C., the deserts bloomed, supporting populations much bigger than exists in the area today. Through this time interval, actually from 4000 to 2000 B.C., the Pastoral Neolithic culture flourished and can be seen in the surviving rock art. Many sites were occupied for periods of months or even years when there were rains. Exactly when the rains fell is unknown. In arid regions, when it occurs, it is restricted to a particular part of the year, so that the livelihood of the inhabitants depends on precise knowledge of when this is. In the eastern Sahara today, there is a division between winter and summer regimes. North of Wadi Halfa, rainfall is in the winter (January to March), but south of it, rains come during summer. There are no rainfalls south of Cairo on a regular yearly basis. This merely reflects the desiccation which began 2000 B.C. Thereafter, it became increasingly difficult to maintain herds of cattle in the desert. The consequent hardship led to large-scale migration of nomads into the Nile Valley, e.g. the invasion of Egypt by the "Sea Peoples", Greeks and Philistines supported by the Libyans, during the reign of Pharaoh Merenptah, son and successor of Pharaoh Rameses II, in the New Kingdom from 1234 to 1220 B.C. This ruler fought against them and also in Palestine, where, for the first time anywhere, the name "Israel" was used in connection with an Egyptian victory. It is on a stele bearing the royal cartouche. A startling increase in rainfall took place in Upper Egypt between 9,000 and 4,500 B.P. due to the Dishnian Pluvial, but there is no agreement as to its origin, its consistency or at what time of the year it fell.

Another critical natural cycle operating since prehistory is the timing of the annual Nile floods and this was established by the Late Pleistocene, i.e. well before the coming of agriculture to Egypt. Floods of Predynastic periods occurred as now, from late June to late September. Neolithic rains are believed to have fallen during January and February. As a punishment for offending the Sun God Ra, the red-haired, red-eyed warrior god Set, murderer of Osiris, having lost his relative infertile southern kingdom, was relegated to the desert borders as the God of Foreigners and the personification of aridity. This might have been the region between the western edge of the cultivated land in the Nile Delta and the Qattara Depression in which occurs a sequence of Miocene to Quaternary beds. The Miocene extends across the northern part of the Western Desert from Wadi Natrun on the east toward north-eastern Libya and continuously outcrops from the Mediterranean to the Siwa Oasis. Sediments range in thickness from 150 to 1,000 m, increasing toward the Nile Delta. They lie unconformably on Paleogene and Cretaceous rocks, including Oligocene gravels and basalt sheets, and are unconformably overlain by the Pliocene. Lateral facies changes are shown from continental, fluviatile and estuarine in the east to open marine in the west. Early Miocene is assigned to the arenaceous and shaly Moghra Formation (75 m thick) of which the lowest member lies directly on the basalts mentioned above. There are some badly preserved marine fossils, but many silicified tree trunks. It is important to note that the Middle Miocene Marmarica Limestone, widely distributed as an extensive plateau to the northwest of the Qattara Depression, is hardly represented in this area (6).

In the Early Pliocene, the Tethys invaded the northern parts of Egypt, the Red Sea and the old Nile Valley so that fluvio-marine and shallow marine deposits accumulated with, to the south, beds of more estuarine facies from freshwater tributaries. The transgression reached its peak in the latest Pliocene when the subsidence of the area stopped and elevation began. Early Pliocene sediments rest unconformably on the Miocene. These are primarily estuarine clays basally with fluvio-marine and shallow marine deposits on the top. The later Pliocene sediments are mainly shallow marine and oolitic limestones with clay beds and a cap of crystalline gypsum 8 m thick which indicates final regression of the sea and subsequent desiccation.

Thereafter, continental conditions prevailed with wet erosional phases alternating with dry climatic conditions. In the former, alluvium was created and structures eroded, while in the latter drift sands, lagoons and duricrusts were formed. Calcareous duricrusts are often conglomeratic and show variable thickness from place to place, originating in semi-arid conditions when wet and dry seasons occurred. It is thought that duricrust formed before the Early Pleistocene degradational process which excavated the main depressions in the area.

Figure 11.1: A soda lake in Wadi Natrun showing sodium carbonate deposits on its desiccated margin.

Along the Nile Valley and the fringe of the Delta, there are several gravel terraces, the highest ones having been deposited by the river at the beginning of the Pleistocene. Later the Nile returned to its old, deeply buried course, reexcavating it and leaving behind successively younger gravel terraces. There are two types of terrace, namely, older ones between 100 m and 200 m above recent sea-level and younger ones constituting terraces lower than 100 m above sea-level. Lagoonal clays and sands are exposed in lowland areas and some canals and there are coastal oolitic ridges (eolean calcarenites) built up along the shore of the retreating sea of the Late Pleistocene. Deposits of sodium salts have been exploited since Pharaonic times from the sabkhas and salty pools in Wadi Natrun which is more than 20 m below sea-level (Figure 11.1).

The entire region is within the mobile shelf of Egypt so that tensional stresses, partly resulting from the subsidence of the Nile Delta, predominated.

Onset of aridity

During the pre-Neogene, widespread transgressions and regressions combined with structural deformation to produce breaks in the record of sedimentation. Such activities stopped during the Neogene and the area was horizontally compressed and later tensionally stressed to produce normal faults.

THE OASIS OF THE FAYUM

Contemporary deposits buried under the arable land of the Nile Valley and Delta also occur in the Fayum Depression which formed in the Early Quaternary. This is separated from the Nile by a water divide lying southwest of Cairo. To the west is the swampy Qarun Lake comprising all that remains of a fourteen times larger ancient Lake Moeris. The Fayum Basin today is 45 m below sea level and the area is 240 km^2. It was created by eolian erosion, tectonic movement and water action and, about 35 Ma ago, in the Early Oligocene, it was the home of Aegyptopethicus.
This low-level adjunct to the Nile Valley incorporates many records of Late Pleistogene events which are otherwise unclear in northern Egypt. The behaviour of its fluctuating lakes is a function of Nile flood-plain levels and occasional sediment blocks at the Nile-Fayum entrance. An aggradation of 34 m of fluviatile deposits near the Fayum Lake represents the Masmas Formation, 20,000 years old.
Several beaches developed along the eastern and southeastern margins of the basin and have artifacts. They can be correlated with the features in the adjacent Nile Valley, thus a beach 34 m above sea-level matches the 8 m fine gravel terrace along the Nile, both possessing Middle Paleolithic implements. About 11,500 B.P., the climate became hyperarid with accompanying total desiccation of the Fayum Lake. At the same time, 20 m of Nile and wadi dissection transpired to a depth of at least 20 m.
About a millenium later, extensive Nile and wadi dissection recurred, this time to a depth of at least 15 m. Aggrading began again and there was a 17 to 20 m water surface in Fayum. By 8,000 B.P., winter rains caused increased flooding of the Nile in Ethiopia, a steeper flood-plain with higher sedimentation rates as well as further flash floods in the wadis. Aggradation recommenced and there were 17 to 24 m lake levels in Fayum.
The lake was partly fed by artesian sources in

addition to local run-off from rains of the Dishnian Pluvial. Difficulties were initially encountered in dating two archeological cultures found in the Fayum and labelled A and B, the former being considered to be older. Radiocarbon dating, not available then, has reversed this order, placing Fayum B between 8,150 and 7,200 B.P., when two early high lake stands (Pre-Moeris and Proto-Moeris) occurred, and the earliest date for Fayum A at 6,400 B.P. This and the Merimda Neolithic are the oldest early farming cultures in the Nile Valley. The subsistence base of these Predynastic peoples probably comprised barley and emmer wheat while they also raised goats, sheep, cattle and pigs. They also indulged in hunting, fishing and plant gathering.

The elevated shorelines of the former lakes are an indication of fluctuating earlier levels. The younger Fayum A is located above the 10 m beach and the older Fayum B between the 10 and 4 m beaches, sometimes dropping down to 2 m. Fayum B sites are usually smaller than Fayum A ones and lack ceramics, having a microlithic technology. The Fayum A culture appears to have been a fully agricultural society, grains of wheat and barley filling many of the sunken silos clustered on high ground overlooking the small villages. Domestic animals, including sheep, cattle and pig occurred. Among others which were hunted, were fish, hippopotamus, elephant, crocodile and carnivores, living in the lacustrine environment. Probably tamarisk and other trees furnished firewood and tools found include millstones, mortars and hammerstones.

From 6500 to 5000 B.P., the climate was semi-arid and wettish and relics of the associated Fayum A and other cultures can be found today. During the time interval 3000 to 2800 B.C., flood levels declined with an accompanying overall reduction in volume of 25-30%. There were flood-plain readjustments and catastrophic low floods between 2250 and 1950 B.C., repeated excessive floods from 1840 to 1770 B.C., declining floods over the half century from 1180 to 1130 B.C. and increasing floods from 600 to 1000 A.D. The Egyptian climate was generally hyper-arid.

However, Lake Moeris still existed and in the 5th century B.C., it was stated by Herodotus to have had a circumference exceeding 700 km and was made during the reign of the Pharaoh Amenemhat I (1980-1970 B.C.). He cut through the Hawara ridge to permit the inflow of Nile river water along the Bahr Yusuf Channel possibly named after the Biblical Joseph. In consequence, the lake enlarged and could have risen

to almost 18 m above sea-level. When the rains ceased and the Nile cut downward, desiccation set in. A temporary reversal was secured when the Pharaoh Amenemhat III (1842 to 1797 B.C.) cleared away stagnant water and converted the Fayum into a verdant area. This was again achieved through the Bahr Yusuf channel which is still in use. Incidentally, this ruler founded a well-equipped center for the exploitation of minerals in Sinai.
The lake then covered more than 1,300 km² and the Oasis remained lush for a while, until in the reigns of the first and second of the Ptolemies (285 to 247 B.C.), it dried up again and was reduced to half its previous extent, thus having to be reclaimed. During the Roman occupation from 30 B.C. to 5 A.D., the lake level dropped drastically from 7 m

Figure 11.2: Lake Qarun with Cardium shell sands along the beach in the Fayum.

below sea-level in the second century to 17 m under this datum in the third. By 1245, when Nabulsi governed the Fayum, the lake reached a level 30 m below the sea. By the time Mohammed Ali came to power between 1805 and 1848, this had declined another 10 m. The lowest recorded level was obtained in 1932 to 1933 when it diminished to less than 46 m below sea-level. Afterwards, it rose again, until in

1978 it was over 2 m higher.

Water leaving the Nile to enter Fayum every year is 10% of its annual volume at Beni Suef from which it can be calculated that four to five years were needed to fill Lake Moeris. The mean depth of the present Lake Qarun is 4 m, the western part being deeper than the eastern. The sediments of the lake are poor in organic contents because of their accelerated decomposition as a result from the high temperatures and the presence of sulfur-reducing bacteria. The tremendous evaporation from this water body results in a level of salinity comparable to that of the open sea. It reaches a maximum exceeding 40‰ in October and it is not surprising that a marine fauna and flora can flourish in it. This arose from the inadvertent introduction of its elements by birds and other agencies. The planktonic flora includes numerous species of diatoms, dinoflagellates, cyanophyceans and chlorophyceans while the zooplankton includes tintinnids, coelenterates and copepods. Among the micro-organisms are thirty three species of foraminifera, mostly bottom-living, and nine species of ostracods. It is noteworthy that the benthonic faunal shells, mostly molluscs like the cockle, often accumulate in elongated mounds along the shores. In the lake itself, there is a variety of salt-water fish, occasionally including the sole (Figure 11.2).

This lake demonstrates the danger of drawing facile conclusions from the sedimentary record in limited areas. Were these fossiliferous deposits, with their clear marine orientation, to be found in a rock sequence, practically all field geologists would probably regard it as representing the onlap of an ancient sea.

THE EVAPORITES OF THE RED SEA

Miocene evaporites in the Gulf of Suez are more than 3,600 m thick and have a major cyclic nature, salinity showing a progressive increase from a normal marine to a saline phase. Some potash salts are recorded, but no separable supersaline phase has been recognized. The salts of the saline phase are succeeded regressively by penesaline sediments of which the top was eroded and unconformably overlain by terrigenous Pliocene and Pleistocene sediments.

Central salt and potash are followed concentrically by other salts until peripherally located anhydrite

Figure 11.3: Section of the taphrogenic sediments in the Gulf of Suez.

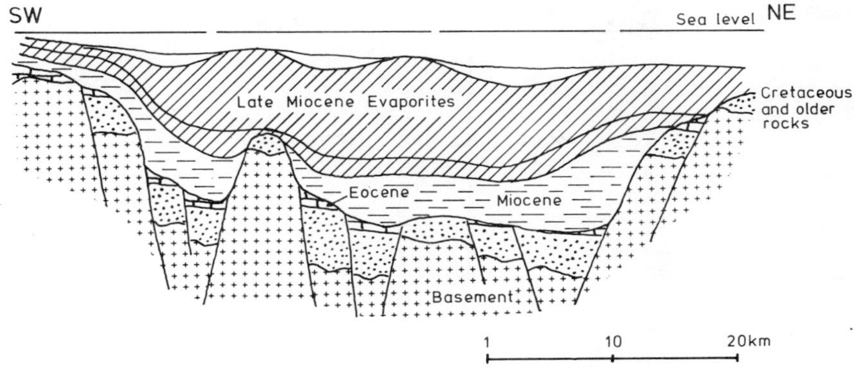

and intercalations of clastics predominate. This sequence was a consequence of Late Oligocene down-faulting of the Gulf of Suez graben and the partial restriction of the depression from the Mediterranean Sea by a forebasinal sill. The various positions of faulted blocks in the trough caused the development of microbasins with salts in the deeper parts and anhydrite marginally. The entire basin appears to be the northerly branch of a larger one involving the Red Sea Depression. As petroleum is associated with evaporites in the Gulf of Suez, it is possible that the conditions of deposition of the evaporites might have been significant in the generation of oil (Figure 11.3).

Gebel el Rousas comprises subsurface basement complex overlain by Middle Miocene followed by Pleistocene to Recent terraces. The Miocene strata are mineralized by hypogene lead-zinc deposits which are anomalous and parallel along zones. The galena which is accumulated in the Middle Miocene is probably of sedimentary origin and also indicated by the relative enrichment in titanium and cadmium coupled with an irregular distribution and low silver, barium and antimony content. It is likely that the mineralization took place near the surface under low pressure at a temperature between 90° and 150° C. In addition, the trace content of hemimorphite, cerussite, wulfenite and other minerals associated with galena demonstrates that a renewed epigenetic hypogene cycle occurred and brought them to the site of deposition (7).

The Miocene rocks in the area between Quseir and

Mersa Alam on the Red Sea coast of Egypt are covered by calcareous crusts (calcrete) which seem to have developed under arid to semi-arid paleoclimatic conditions. Associated karst features indicate that the Miocene outcrops of the Gebel El Rusas and Abu Dabbab Formations were exposed to repeated cycles of precipitation and evaporation during the Pliocene or Pleistocene. In addition, the Abu Dabbab Formation, is evaporitic, composed of anhydrite, gypsum and some dolostones. It is capped by the "oil-tainted" limestone (8).

The Miocene, which is several hundred meters thick, comprises conglomerates, sandstones, sandy clays and, toward the top fossiliferous limestones and sandy shales. In the latest Miocene, the Messinian, up to 40 m of gypsum were deposited.

DESICCATED LEVANT

The Lisan evaporites of the Dead Sea are restricted to the central and structurally deepest zone of the Rift province. Seismic reflection data indicate a thickness of at least 4,000 m and hence they comprise almost half of the total post-Proterozoic sedimentary column beneath this Peninsula. The evaporites are mostly halite with some thin carnallite layers together with some anhydrite, shale and marl. The shale and marl thicken rapidly to the east and more gradually to the south. It appears that 20 km south of the Dead Sea, the salt mass is practically totally replaced by a well-stratified sequence of shale and marl. The salt most likely represents a gentle doming and palynologic investigations demonstrated that it is maybe Late Miocene, certainly Pliocene and possibly Early Pleistocene in age.

The nearby Mount Sodom diapir is of Pliocene-Pleistocene age. It is a vertically layered salt wall extending meridionally more than 2 km parallel to the strike of the layers. Its summit is transected by a dissolution salt table overlain by a residual caprock. The deformation features include rotated boudins, stretched desiccation polygons and fracture cleavage. These phenomena derived from salt flowage. Strain is greatest along the margin of the diapir which has an annual rate of rise of 3-4 mm possibly since the Late Pleistocene (9).

Along the eastern side of the Jordan River valley, there are well-cemented and coarse-grained clastics which alternate with travertines and overlie older rocks unconformably. These 100 m thick clastics belong to the Shagur Formation which have been strongly structurally deformed. They are continental deposits and contain gastropods and ostracods of Late Pliocene to Early Pleistocene age. They represent the oldest Quaternary rocks in Jordan. In gravels on this same side of the river, Olduwan-type pebble tools have been found and the beds were assigned to the Middle Pleistocene.

In the central Jordan River valley, there is a sequence of alternating conglomerate, sandstone, marl, shale and a red paleosol. This is more than 350 m thick and also structurally deformed to a lesser extent than the Shagur Formation. Its age is tentatively given as Early Pleistocene. All older rock units in the rift province are unconformably overlain by the Lisan Formation. This comprises mostly shale and marl with an upper varvitic shale with gypsum and sulfur. It is of Late Pleistocene age and was deposited in a fluctuating oligo- and miohaline lacustrine environment. Chert implements have been found in its upper parts. At its maximum size, the Lisan Lake covered the whole Rift Valley from the Sea of Galilee to 80 km south of the Dead Sea. It was almost 300 km long (10).

THE ARABIAN PENINSULA

Evaporitic beds associated with intra-formational breaks and sometimes diapiric occur in the Callovian to Tithonian part of the Late Jurassic in Arabia Felix deposited on an irregular submarine basement topography. In eastern South Yemen, salt domes contain rock salt with layered structure and stains of bitumen. Since the overlying shales dip conformably with the evaporitic layers, it is believed that there is a stratigraphic relationship, the salt being depositional and non-diapiric. Elsewhere, surface salt domes have been recognized. However, no other such surface manifestations of Jurassic diapirs occur anywhere in this region. Fossils from underlying limestones and clastics of the same Madbi Formation may be of Oxfordian age, but definite Kimmeridgian ones have been identified and even some Early Tithonian as well. This formation ranges up to 275 m in thickness. Evaporites were also deposited during the commencement of the overlying Naifa Formation, although most of this was an open sea facies. The more northerly lagoonal environment of deposition and an associated barrier isolated the inland salt dome areas. Alternatively, the rise of evaporitic conditions were a result of stagnation on a submarine plateau where rapid evaporation became possible through the absence of a deep-water current circulation. A combination of both mechanisms cannot be excluded (11).

In Kuwait, the end of the Jurassic may be the top of the first anhydrite bed below Early Cretaceous rocks, this being perhaps equivalent to the Hith Anhydrite of Saudi Arabia. From the deep wells in the Burgan Oil Field, 1,350 m of Jurassic rock were confirmed and the uppermost 450 m of this is made up of evaporites. The lower 900 m comprises mostly limestones, but there are thin anhydrite beds in it as well. Underlying all this, there are recorded to the base of the borehole 200 m of Triassic clay, marl, shale and sandstone. The Late Jurassic evaporites are well-known from the Zagros Mountains but they also extend into the Palmyrian Mountains. No equivalent salt deposits have been found in Jordan and Egypt, although in the former there are gypsiferous shales in the Late Cretaceous.

There is probably a hiatus between Cretaceous and Paleocene in Saudi Arabia preceding a widespread Danian transgression which deposited thick neritic limestones and basinal marls. Carbonate rocks extend into Iran and Iraq, most of the latter being covered by globigerinal marl. Deposits of a flysch-type accumulated in the trough. The Early Eocene ushered in persistent and extensive evaporite precipitation so that great thicknesses of anhydrite were deposited in the 55 m thick Rus Formation over the Rub al Khali Basin as well as in South Yemen, across Qatar, the western Persian Gulf, northeastern Arabia, Kuwait and southern Iraq. In Bahrain, the Rus Formation is the oldest which outcrops, is located in the central part of the flat anticlinal island and reaches a total thickness of 65 m. Made up of chalky and cherty dolomitic limestones, it includes anhydrite beds and quartz geodes of which some contain petroleum. However, the oil-producing horizons are in the Middle Cretaceous. The Rus Formation is called the "Zone C aquifer" and yields a limited amount of fresh water derived from rainfall, this floating on top of the usually saline water of the Formation to comprise a cap up to 4.5 m thick (12).

Although normal marine conditions existed in adjacent regions, there was red-bed deposition along the northeastern flank of the still subsiding Iraq-Iran Trough. This arid Paleogene episode must be carefully distinguished from the desiccation of the Mediterranean and its accompanying phenomena in the Nile Valley and the Levant commonly referred to as the Miocene Crisis. Fresher seawater invaded in the Middle Eocene and carbonates, both of nummulitic and globigerinal facies, deposited. The pre-evaporitic type of sedimentation pattern prevailed yet again.

The Shield emerged in later Middle Eocene times and reduced the size of the Tethys, this situation persisting until now. Thus, except for occasional minor coastal flooding in the Middle Miocene, continental conditions have predominated. The Miocene in Arabia is really a thin wedge of lacustrine, fluvial and coastal plain deposits marginal to the principal area of subsidence in Iran and Iraq where evaporitic conditions obtained. In these latter countries, marine sedimentation went on for a short time after a Lower Fars evaporite phase.

For instance, in Iraq the heterogeneous Lower Fars Formation is the basal and evaporitic part of the tripartite Fars Series. The Early Fars comprises repeated cycles of the sequence: marl, limestone and gypsum. An intertidal depositional environment for the stromatolites is evidenced by the characteristics of the cryptalgal limestones and also by the presence of calcite pseudomorphs after gypsum. It is concluded that the sulfate horizons formed in a supratidal environment, the cycles resulting from repeated sabkha progradation. Otherwise,the Lower Fars fossils include oysters, pectens and microfossils.

The Middle Fars is almost missing outside Iran, hence is usually included as an Upper Fars division. Lower Fars is then restricted to the evaporitic Miocene below the highest bedded anhydrites. Westward from the Euphrates, it becomes less and less evaporitic and grades into clastics both laterally and vertically. However, for the most part the sequence comprises limestones and mudstones with gypsum. The evaporitic phase was closed by large scale deformation. The rising mountain folds contributed masses of clastics into adjacent synclines. Initially, there were red-beds of the Upper Fars and then, following more rapid deformation, coarse sands and conglomerates. These two sets of clastics reached a maximum thickness of 4 km, paralleling the Nile Delta (13).

Despite intense diastrophism across the Persian Gulf, there occured very little tectonic activity in

Arabia. Rocks of presumed Pliocene age dip only 1 or 2 m per km toward the Persian Gulf, this probably representing tectonic tilting activated by the rather remote subduction zone below the Zagros Mountains to the east. Notwithstanding, there are vast lava flows in the Shield and northwestern Arabia which attest to considerable Pliocene and later volcanism. In the Quaternary, the great sand deserts and sheet gravels developed to provide the present aspect of Arabia Deserta.

Calcareous duricrust indicates semi-arid conditions and may provide evidence of climatic changes. In eastern Saudi Arabia it blankets a rugged Neogene topography and exhibits various stages of maturity. The least indurated crust rests on the youngest surfaces which developed recently and vice versa. Thicker and more mature crust rests on the oldest relief, so that the thickness and degree of coherence of any crust may be taken as an indication of its relative age.

The forming of calcretes involves recrystallization of original calcite in parental sandy limestone of Miocene and Pliocene age, addition of secondary calcite from above or below and replacement of quartz and feldspar by calcite. The calcium carbonate content of any particular crust is directly related to its age. Duricrust progressively and sporadically formed during semi-arid episodes of the Late Pliocene, Pleistocene and Early Holocene, thus acting as a paleoclimatic indicator.

REFERENCES

1. HSU, K.J., MONTADERT, L., BERNOULLI, D., 1977. History of the Mediterranean salinity crisis. Nature, 267, 399-403.

2. El-HAWAT, A.S., 1980. Intertidal and storm sedimentation from Wadi al Qattarah Member, Ar Rajmah Formation (Middle Miocene), Al Jabal al Akhdar. In: The Geology of Libya, 2, ed. SALEM, M.J., BUSREWIL, M.T., 449-461, Acad. Press, London, UK.

3. BUTZER, K.W., 1980. Pleistocene history of the Nile Valley in Egypt and Lower Nubia. In: The Sahara and the Nile, ed. by WILLIAMS, M.A.J., FAURE, H., 253-280, Balkema, Rotterdam, NL.

4. HOFFMAN, M.A., 1980. Egypt before the Pharaohs. Routledge and Kegan Paul, 391 pp., London, UK.

5. SMITH, A.B., 1980. The Neolithic tradition in the Sahara. In: The Sahara and the Nile, ed. WILLIAMS, M.A.J., FAURE, H., 451-465, Balkema, Rotterdam, NL.

6. OMARA, S.M., SANAD, S., 1975. Rock stratigraphy and structural features of the area between Wadi El Natrun and the Moghra Depression (Western Desert, Egypt). Geol. Jb., B 16, 45-73.

7. RASMY, M., 1981. Trace-elements content of galenas and associated minerals in some Miocene lead-zinc deposits near Red Sea coast, Egypt. Geol. Rdsch., 70, 874-881.

8. El AREF, M., WAHAB, S.,A. and AHMED, S., 1985. Surficial calcareous crust of caliche type along the Red Sea coast, Egypt. Geol. Rdsch., 74, 155-163.

9. ZAK, I., Freund, R., 1980. Strain measurements in eastern marginal shear zone of Mount Sedom salt diapir, Israel. AAPG, Bull., 64, 568-581.

10. BENDER, F., 1975. Geology of the Arabian Peninsula, Jordan. USGS Prof. Pap., 560-I, 101-136.

11. POWERS, R.W., RAMIREZ, L.F., REDMOND, C.D. and ELBERG, E.L. Jr., 1966, Geology of the Arabian Peninsula - Sedimentary Geology of Saudi Arabia. USGS Prof. Pap., 560-D, 1-147.

12. GREENWOOD, J.E.G.W., BLEACKLEY, D., 1967. Geology of the Arabian Peninsula - Aden Protectorate, 560-C, 1-96.

13. SHAWKAT, M.G., TUCKER, M.E., 1978. Stromatolites and sabkha cycles from the Lower Fars Formation (Miocene) of Iraq. Geol. Rdsch., 67, 1-14.

Part V

THE DRAMA OF CLIMATIC CHANGE

Aridity in northeastern Africa and Arabian Peninsula was perennial after the Miocene, but during the last two million years the latest ice age occurred. The higher mountains of the region became ice-capped, the snowline was depressed, cirques were excavated and some moraines deposited. A number of pluvial periods were associated with interglacial stages. Marine advances and retreats left terraces around the Mediterranean. Coastal dunes formed. Fossil coral reefs in the Red Sea attest to warm episodes and their affinities confirm an opening to the Indian Ocean. Cultural influences like the Acheulian and Mousterian left fascinating traces such as rock art. The Libyan Desert silica glass was occasionally fashioned into tools by these peoples. Recent investigations show it to be of sedimentary, not extraterrestrial, origin.

The Sahara Desert, the driest place on Earth, is a deflated area of barchans, wind-abraded ridge-and-swale systems, wadis and yardangs with a precipitation under 100 mm annually. Its southwest Egyptian part has been compared with Mars. The satellite imaging radar of the space shuttle "Columbia" in 1981 penetrated several meters into its sands to reveal a buried network of "radar rivers". Their flowlines are inconsistent with the present regional trend of the water table. Evaporation potentials certainly reach 5,000 mm per year or more, so the aquifers are composed mainly of fossil water. In Libya, similar arid conditions exist, but the Romans nevertheless colonized the barren country and built many dams after 69 A.D. A Neolithic way of life began under cold, wet climatic conditions during the seventh to fifth millenia B.P., becoming confined to oases by the third millenium. This was linked with a major climatic change and followed by aridity until the "Little Ice Age" restored Neolithic conditions in Africa.

Chapter V-12

ICE AGE IMPACT

"A gauntlet of ice which centuries ago Winter threw down in defiance to the Sun".
Henry Wadsworth Longfellow, Journal, 16 July 1836.

PERIGLACIAL PHENOMENA

During the last Ice Age, high mountains between the Mediterranean and the arid zone may have been covered with sizeable glaciers and the snowline in the Lebanon Mountains and at Mount Hermon could have been sometimes depressed as much as 900 m below the summits. A rather convincing indication of glacial activity in the region is provided by the cirque-like valley on the humid western slopes of the mountains near to the Valley of the Cedars. In its mouth, there is a crescentic moraine-like deposit at 2,000 m made up of tufaceous limestone bearing plant imprints mixed within a gravel breccia. Unfortunately, no striated rocks have been found in it.
In the Hoggar, there is a cryonivational zone of slow mass movement at 2,200 and 2,400 m, a nivational zone on the east-facing slopes at 2,400 and 2,600 m. Some forms may indicate small glaciers on the summits of Assekrem and Tahat. Around the northern peaks of the Tibesti, nivational forms can be seen only above 3,000 m. The 2 km long Muskorbé Massif is the highest of these at 3,376 m, constituting remnants of an earlier glacial episode. A further evidence for this has been provided from the Simen Mountains of Ethiopia in the form of terminal moraines down to 2,600 m in the Mai Shaha Valley. Strangely, south of the Tibesti at the 3,415 m high Emi Koussi, they are not present although this is the highest point in the Sahara. This difference in the level of snow-patch erosional activity indicates important climatic differences in the Sahara at this

time (1).

Periglacial processes are still active today in suitably exposed localities above 2,500 m in Lebanon and on Mount Hermon where patterned ground is found. They impede solifluxion at lower levels. Contemporary periglacial activity vanishes in the Sinai, although during the last major cold phase periglacial landforms are very evident on Precambrian basement near Gebel Catharina. In addition, a cirque-like form with a northeastern exposure at 2,600 m on Mount Sinai could have been the site of an ice-field during the Pleistocene. However, as in the Valley of the Cedars, there are again no striated rocks.

Despite many days of recurrent frost in Tibesti and the Hoggar, there is no sign of regular and morphodynamically effective periglacial processes there. The reason for this is the lack of ground moisture during the frost period because most precipitation falls during the warmest part of the year. In the Saharan uplands, an ectropical seasonal climate with winter frosts encounters a tropical monsoonal rainfall pattern of summer rains. The consequence is that freeze-thaw processes are ineffective through lack of moisture. Additionally, not only would an active periglacial zone be very high, but in a hyper-arid climate it would just disappear. Winter frost is far less important as an erosional and transporting agent than are the recurring heavy summer rains in the mountains.

As regards the fluviatile piedmont deposits of the region, they demonstrate three Late Pleistocene episodes. The earliest is an aggradational event from which a widespread geomorphic crisis of climatic origin has been recognized. An interval of extreme aridity is demonstrated in the Upper Terrace in northwestern Tibesti which accumulated during a phase of climatic desiccation and extended north into the Yebbiguè Basin. Here, a red-brown paleosol possibly indicates preceding pedogenesis during one of the pluvials. The sediments of the Upper Terrace are rhexistatic. A silty and fossiliferous Middle Terrace occurs in the Bardiguè catchment of northwestern Tibesti, which according to radiocarbon data, was probably deposited between 15,000 and 8,000 B.P under rainfall condition not unlike those of present-day tropical storms. A more recent terrace of similar kind contains freshwater molluscs dated at 6,000 B.P., i.e. of Middle Holocene, probably Dishnian Pluvial age.

THE EASTERN MEDITERRANEAN

The elevated marine terraces of the countries of the Levante correspond with the dramatic interglacial and glacial stages of the Pleistogene in Europe. Since they are widely distributed and occupy roughly the same elevations throughout the region, Lebanon may be taken as typical. This is advantageous because of its diverse geomorphology and better known archeology. The mountain chain of the Syrian Arc rises abruptly from the drowned coast and former high sea-levels have left their mark in the shape of raised beaches and terraces. This profile of abandoned terraces is perhaps the best preserved in the Eastern Mediterranean.

One terrace at 100 m elevation extends for nearly one-third of the 220 km length of the coastline, testifying to the Sicilian Transgression which ushered in the Ice Age. This ancient shore was succeeded by a regression to 30 m. Coastal dunes formed after the deflation of littoral sediments. Earliest Paleolithic implements have been found in this terrace. Another major retreat of the sea down to the existing level was accompanied by a second widespread development of coastal dunes. Following the subsequent Milazzian transgression, a new terrace was deposited at 45 m which contains Middle Acheulian and Tayacian tools. Thereafter, sea-level dropped more than 6 m. These Tayacian and Levalloisian cultural relics may well confirm the correlation of this ultimate regression with the Mindel Glaciation. As well as the eustatic movements, the mobile epicontinental margin affected the level of these Early Pleistocene terraces.

The first undisturbed terrace is at 35 m elevation and represents the Tyrrhenian Transgression. It preceded another marine retreat to below the present sea-level probably an effect of the Riss Glaciation. Younger Monastirian beaches at 15 and 6 m contain Levalloisian and Mousterian artifacts as well as the thermophilic gastropod Strombus bubonius. This certainly represents the Eem Interglacial. After this warm interval, pluvial deposits mark the onset of cold and wet conditions prior to the Würm Regression to well below modern sea-level. At this time, the average water temperature in the Eastern Mediterranean was appreciably colder than now. The Flandrian Transgression lasted from the Neolithic to the Predynastic Age and elevated the Mediterranean to 4 m above its existing level.

This last transgression is responsible for the Nizza Terrace and may be represented in Egypt by a long offshore bar of oolitic limestone. A 2 m high terrace has been cut in this. Also, there is a former submarine shelf extending 4 or 5 m inland and reaching another 2 m terrace. This may correspond to the Nizza Terrace. Beyond this is another former submarine shelf stretching 10 m inland. Almost at sea-level, behind earlier bars, there are dry lagoon beds which are covered with terra rossa. The submergent coastline of today drowned Greek and Roman classical buildings which is also the case in Israel and the Nile Delta. This marine invasion is not mainly due to eustacy, but rather to a tectonic downwarping.

The Israeli section of the Mediterranean coastal plain comprises two main parts. There is a low lying coastal feature which started to subside before the Neogene and is still doing so. In addition, there is mountainous coast on which the Pleistocene shore-lines are recorded in marine terraces of raised beaches as at Carmel. There were three Quaternary transgressions, the Milazzian, the Tyrrhenian and the Monastirian. Succeeding them is the Flandrian Transgression which started in the Holocene and is still continuing. The maximal encroachment was that of the Milazzian and the minimal that of the Monastirian. The faunal analysis indicates that the sea-level of each transgression was lower than that of its predecessor. Pleistocene shore-lines are generally parallel, but do not match the present-day one.

RED SEA CORAL REEFS

Fossil fringing reefs as old as the Miocene have been recognized on the shores of the Red Sea since the beginning of this century. Of course, it had a Tethyan fauna before opening to the Indian Ocean in the Pliocene. The later biostromes are taken to be Quaternary because their coral and molluscan assemblages are Indo-Pacific in type and entererd through the Bab al Mandab Straits. It seems probable that these raised reef platforms reflect higher sea-levels during the Pleistogene despite crustal mobility in this region, which includes localized uplifts. Such old aragonitic skeletons must have undergone diagenesis through percolating water, although this takes place later than the conversion of aragonitic cements into calcite. Interestingly, the diagenesis of aragonitic mollusc shells in Red Sea reefs occurs after that of corals and cements. These effects are attributable to several factors. One is the different influences of various organic substances during aragonite precipitation. Another is linked to the density of the crystal fabric. Furthermore the degree of saturation of the meteoric water with regard to calcium carbonate within the diagenetic environment is critical (2).

Figure 12.1: Stone Age workshop in the desert near Wadi Howar, Sudan.

When the Late Paleolithic culture flourished, 15,000 B.P., the sea-level was minimal at 120 m below its present elevation. During the Dishnian Pluvial, it rose until fringing reefs expanded from 7,000 to 5,000 years ago. Radioisotope dates imply that as early as 30,000 years B.P., the shorelines of the Red Sea paralleled those of today. Old reef limestones from the northern Farasan Shelf gave ages of 35,000 to 24,000 years. However, still earlier reef limestones now 10 m above sea level produced ages of 90,000 to 70,000 years. Along the Red Sea, therefore, Late Pleistogene fringing reefs of varying ages grew randomly together. In the Gulf of Aqaba at Elat, three raised beach terraces with coral reefs are clearly identifiable. At 250,000 to 100,000 years of age, they are the oldest such features in the region. Apart from these which are 200 m high, the other terraces are at about the same elevation as their Mediterranean counterparts (Figure 12.1).

The western Red Sea littoral, near Mersa Alam, contains tectonically undisturbed high beaches. The lowest is a coral reef at 10 m elevation possibly representing the early Last Interglacial high-stand. Another coral reef is 6 m higher and has been dated by thorium-230/uranium-234 as 80,000 B.P., i.e. mid-last Interglacial transgression. Both episodes were accompanied by minimal flash flooding in the wadi. Older gravel bars and estuarine gravel accumulated and are now 25 m high. They indicate a prior acceleration in wadi activity. Before this, there was regressive oscillation and an earlier alluviation of coarse gravels of the Middle Terrace showing pluvial conditions associated with the last Glacial regression. Preceding all of these events, there was a dissection of alluvium and calcification inferred from a paleosol. An alluviation of the Lower Terrace again indicates pluvial conditions of the early or middle parts of the Last Glacial regression. Long-term dissection of coastal deposits to well below modern sea-level started toward the end of the Last Glacial regression. All this shows that the terminal Pleistocene and Holocene fluctuations of moisture were too insignificant to have any geomorphic effect on the hyper-arid Red Sea littoral.

The unusual climatic conditions there today continue to encourage widespread development of such fringing reefs grown by forty genera of living corals. They vary considerably in character depending on where they grow. This can be on the coast or in calm water bays and occasionally as barriers far out to sea as, for instance, the Wingate Reef 5 km off Port Sudan. To the south, an interesting case of a living reef with an overlying fossil one may be seen fringing and almost closing the mouth of the harbor of the old Sudanese town of Suakin, formerly an infamous commercial center of the slave trade. As the entrance has been reduced to a very narrow channel and pilgrimages to Mecca are mostly made by air, there is practically no remaining mercantilism. This focal city of Islam has a flourishing port, Jeddah, where a stable sea-level has stimulated very rich coral growth and the building of extensive reef plates. On the other side of the Sea, at Flamingo Bay near Port Sudan, plateau-like reefs occur in a system of outer and inner features separated by deep channels and lagoons. The situation is completely different from that in Sri Lanka and many areas of the Pacific where a normal type of lagoon reef manifests a rising sea-level. This may appear strange because these regions fall into one of the two evolutionary centers of hermatypic corals in the oceans, viz. the Indo-Pacific one. However, it is attributable to different stages in the geologic development of Flamingo Bay. The other evolutionary center is that of the Atlantic.

In fringing reefs, a shore with beach rock, pebbles and sands having weak rip currents and spray is succeeded by a calm water lagoon and marginal to the latter is a reef platform. The seaward edge is highly turbulent with sizeable surf and enormous colonies of the fire coral Millepora associated with encrusting algae. The succeeding slope is steep and composed of dead coral rock overgrown by a thicket of living corals including Acropora at the top and Lobophyllia below. Finally come the fore reef with brain corals and the littoral flat which reaches a maximum depth of 40 m. Skeletal carbonates of such recent and fossil reef builders appear to be labelled isotopically. They have characteristic heavy stable isotope ratios which vary as a function of their recent or paleo-environmental distributions as recorded in the water in which they are living or by rock facies.

THE PERSIAN GULF

The Persian Gulf is an extremely shallow, unusually saline sea with an average depth of only 35 m and a maximum of 100 m, so that its floor comprises part of the continental shelf. Surface water temperatures in the coastal regions range from 10°C to 35 C° in summer and far off-shore the range is from 15 C° to 33°C. There is very low rainfall and the surrounding lands are very arid. it is not surprising that the evaporative loss of water from the Gulf far exceeds input from rivers and run-off. This is manifested by a high salinity which is among the most important environmental factors governing the occurrence and distribution of marine life. Since the Gulf is connected with the neighboring Indian Ocean by the narrow passage of the Strait of Hormuz, its high salinities and wide temperature fluctuations are not appreciably dampened by exchange of waters. In fact, the Gulf has a Mediterranean circulation pattern involving emission of salty water through the bottom of this Strait, a compensating quantity of fresher, lighter Indian Ocean water penetrating inward at the surface. The resultant is so modest that the steady state conditions of the Gulf persist in differing markedly from those of the Indian Ocean. It was recently added to the seas by crustal movements following the separation of the Arabian Plate from the African continent which involved the creation of a gap now filled by the Red Sea. The other edge of this plate rammed Asia to form the Zagros Mountains along the collisional contact, where gentle down-warping produced the floor of the existing Gulf when it dropped below sea-level.

This easterly movement of the Arabian Plate began in the Miocene, with the Gulf itself forming in the Late Pliocene 3 to 4 Ma ago. Raised beaches in Oman up to 350 m date back to the latter. During each of the subsequent Pleistocene glacial phases, lowerings of sea-level occurred to a maximum extent of 120 m. Accordingly, the Gulf Basin was drained to become a wide, shallow valley containing the combined flows of the Tigris and Euphrates Rivers. They reached the sea where the Strait of Hormuz is today. Thus, the Gulf was a mere estuary several times in its rather short history. The sea cut a series of platforms at the stable level of maximum retreat and at periods of relative standstill during the post-glacial rise. The altimetric sequence of terraces in the Gulf at Kharg Island comprises levels matching those of the Mediterranean up to the 60 m one which corresponds to the Milazzian Transgression.

In connection with the Pleistogene invasions or retreats of the sea, a karstic hydrology developed. This characterizes Bahrain and the phenomenon was already observed by Ibn Battuta in the 14th Century A.D. He recorded that,during the Turkish occupation, the sailors used to dive into the waters of the gulf, bringing up potable water in leather bags for the use of their captain. Also, the Portugese supplied themselves in the same way by means of pumps. There is an apocryphal story that a camel once fell into a spring, only to reappear later in the sea.

Despite Pleistocene physiographic modifications, the fundamental tectonic control remains apparent and there is a partial correlation between bathymetric and structural highs and lows. The type of marine sediment deposited is mainly a function of those biological communities providing skeletal material and these vary in nature from shoals to depressions. Recent unconsolidated sediments result from the post-glacial Flandrian Transgression beginning about

18,000 years ago and attaining its present level about 5,000 years ago. The thickest recent sediments are found along the axis of the Gulf. Could a folk memory of this event be preserved in the record of Genesis, 7, referring to the Deluge when "the waters prevailed exceedingly upon the Earth: and all the high hills, that were under the whole heaven were covered. Fifteen cubits upward did the waters prevail: and the mountains were covered". The cubit is taken to be 0.4632 m in the Greek or Olympic measuring scale and 0.525 m in that of the Ancient Egyptians.

In Kuwait Bay, the carbonate content of the recent sediments has a triple origin, i.e. from dust storm fallout deposits over the country, material carried in suspension by currents from Shatt Al-Arab and skeletal parts of micro-organisms which may originate in the sediments. The carbonate content increases in the coarse rather than in the fine sediments (3).

The Gulf is a basin of Late Pliocene to Pleistocene age with tectonically-influenced morphology. The relief is more subdued on the Arabian side because of lesser disturbance and folding, faulting and salt diapirism are superimposed upon mainly meridional "Arabian faults". By contrast, the intense folding of the Zagros Orogeny controls the topography on the Iranian side. It is apparent that interference occurred between Arabian and Zagros folds. The tectonically governed-morphological pattern was suppressed by Pleistocene limestone sedimentation, although local rejuvenation resulting from the Quaternary tectonic adjustments did take place.

NILOTIC EGYPT

Quaternary sediments in Egypt cover all preceding deposits unconformably and are thickest in the Nile Valley. They accumulated during several pluvial and interpluvial episodes of which the former apparently correlate with worldwide rises of sea-level and warmer climates. The pluvials are inferred from spring, torrential and lacustrine deposits as well as paleosols. Interpluvials are represented by inland dunes, evaporites and sabkhas.

Early Quaternary deposits in the Nile Valley are of local provenance, actually the products of the first Quaternary pluvial, the Idfuan. Sediments of younger Nile stages derive from sources outside and south of

Egypt. The Plio-Pleistocene boundary is placed at 1.85 Ma B.P. at the Olduvai magnetic interval by some and at 2.8 Ma B.P. at the Kaena magnetic interval by others. Subsequently, there are at least eight major pluvials of which the first pair is the Idfuan and the Armantian of Early Quaternary age. Their position is rather ambiguous because they are neither preceded nor succeeded by interpluvial deposits. The six younger ones are the Abbassian, the three Saharan stages , the Kubbaniyan and the Dishnian. The effects of these pluvials may have been confined to the southern reaches of Egypt, the country having been subjected to two different climatic models during the Quaternary. The southern one represented a migration north of the Sudano-Sahelian savanna belt and the northern a Mediterranean pre-Sahara steppe. The younger pluvials may be tentatively correlated with interglacial stages in Europe and North America. The Abbassian correlates with the Eem or Sangamonian Interglacial. Subsequent pluvials may be linked with interstades of the Würm or Wisconsinan Glaciation (4,5).

The Kom Ombo plain is in the Nile Valley where faults in the Nubian sandstones created a small graben crossing the river. It is about 40 km north of Aswan, covers an area of 400 km^2 and lies mainly on the east bank of the Nile. Its surface is composed of Late Pleistocene interbedded sands and silts deposited either by the river or by wadis in the east. Drainage from these wadis has laid down silts as deltas or fans which have forced the Nile to the eastern side of the graben. Pleistocene vertebrate fossils have been recorded and include a variety of mammals such as hippopotamus, cattle, gazelle, spotted hyena and hare together with water birds and fish. They come from sites in the Gebel Silsila Formation. These are in the Younger Channel silts accumulated in abundant channels of earlier stages of the Nile. They have been radiocarbon-dated as between 17,000 and 12,500 B.P., i.e. mostly predating the Dishnian Pluvial (6).

Evidence from these and other faunas indicates that an exchange went on over the Suez Isthmus during the Late Pleistocene and Early Holocene. Among new immigrants into Africa were wild cattle, probably during the Villafranchian, the ass, possibly during the early part of the Late Pleistocene and the pig, perhaps introduced by Man after 12,000 B.P. The wild cattle may have displaced the African buffalos. The spotted hyena is no longer in Egypt, having been replaced by the striped hyena which, like its cousin, first entered Africa during the Early Pleistocene. It may be inferred that the Suez Isthmus, still an effective bridge between the African and Eurasian faunas as late as the Early Holocene, originally became important as a mammalian migratory route during the Miocene. Probably only Man has prevented continuance of this trek through settlement and accompanying agriculture.

The fauna shows that species existed which cultivated savanna, scrubland, rivers and lakes. The wild cattle and buffalos used open forest woodland and needed extensive water for cooling purposes. Hippopotami are always near permanent water. However, the possible presence of white gazelle and Barbary sheep imply a drier environment, suggesting that elements from surrounding hills are represented. Catfish remains indicate the existence of lasting meanders or channel pools and shallows of which the maintenance would necessitate a more or less constant flow of water throughout the year with seasonal fluctuation to flood lower lying areas. Additional water would have been supplied by rainfall. Water birds attest to shallow lakes and pools which they would have needed to find their food in, although some, e.g. geese, forage on dry land. It is interesting to note that the mammals are often represented by immature individuals which suggests that the sites were probably inhabited through most of the year.

THE EASTERN SAHARA

The Western Desert is the driest part of Egypt and covers about two-thirds of the country. It comprises a vast plateau of moderate elevation under 500 m above sea-level with sand dunes over about 40% of it and there are inselbergs and cuestas as well. Great thicknesses of mostly undisturbed sedimentary rocks have been deposited. Sandstones in the south grade into limestones in the middle which mix with calcified sands in the north. Uniformity of surface together with aridity facilitates eolian transport of sand and increases the erosional capability of both. This is why conspicuous relief is absent. However, some domal mountains occur because their sedimentary cover weathered away and igneous rock became exposed as at Gebel Uweinat. The erosional surfaces are practically desert peneplanes. In fact, this marginal part of Gondwana has been exposed to frequent peneplanation. This stage is now being approached again in contrast to the situation in the still youthful Eastern Desert.

During the Early Holocene, a gradient of decreasing rainfall from west to east developed along the Tropic of Cancer. Pleistogene sediments contain floristic elements pointing to the existence of a dense Mediterranean forest cover in the Saharan Mountains during pluvials. Temperate elements occurred as far south as the northern Tibesti and pollen of Mediterranean plants are equally widespread showing long-distance transport. The flora is modern in character. In the mountain areas, a dry Mediterranean open woodland with appropriate species is to be expected, but no forests would have occurred in the upper part and the lower regions must have had a gallery-like vegetation with Acacia, Tamarix, etc.

In the Western Desert, there are silty clays deposited in stagnant waters of ephemeral lakes with low organic and negligible carbonate contents. However, in the west there are coeval calcitic lake sediments deposited in a drainage system which extended to the Mediterranean. It may be noted that a meridional south to north gradient of diminishing rainfall also occurs. In the northern Libyan Desert, from west of Kharga to the eastern edge of the Kufra depression, silty pelites predominate. However, in the adjacent area to the south around Selima and as far as Wadi Howar, a carbonate-limnic facies is found.

In the Libyan Desert, pelitic sediments fall into two categories. Firstly, they can be of mountain- or Gilf Kebir-type comprising interbedded sand and gravel or fluvially transported drift sand. Secondly, they can be plains-type with the interbedded material consisting only of drift sand. These imply a semi-arid climate with severe wind erosion and precipitation under 200 mm annually. In the Great Sand Sea, carbonates and pelites formed in shallow lakes and calcretes are found. Pelite sedimentation began after 11,000 B.P. Up to about 26,000 B.P., there was a hyper-arid climate with deflation and eolian accumulation. Holocene pelites are due to local rainfall.

Other than in the Nile Valley proper, the soils of southwestern Egypt are almost devoid of humus. However, eolites, carbonates, occasional sulfatic and alkalinic soils are encountered. Beneath the desert pavement or dune sand, there is a thin, but dense silt layer as well as an extremely porous horizon called "Schaumboden" of columnar structure filled with wind-blown sand in the cracks. Wet and saline soils are mostly those which are irrigated in the oases and by naturally occurring groundwater and the subsurface moisture. From the distribution of hitching stones, it is apparent that sometimes during the Early Holocene, livestock, most likely cattle, could have been reared from the Great Sand Sea to the Wadi Howar. Separation of vegetated areas was sufficiently short to allow the migration of large mammals such as giraffes and elephants into the Gilf Kebir. In the Middle Holocene, the Wadi Howar periodically debouched into the Nile and, during dry episodes, developed dunes (7.8).

DESERT DEPRESSIONS AND LAKES

One of the main features of the Western Desert is its internal drainage system, the interior basins manifesting the tremendous influence of deflation in sculpting and deepening depressions even below sea-level. Such topographic lows are located on or near significant geological boundaries. Some take the view that they are tectonic in origin, others that they were hollowed out in an area of domes and yet others that they resulted from the influence of running water. It was long ago suggested that the Kharga Depression began to be excavated in the Pleistocene by the erosional action of running water. This view is no longer considered to be valid.

The depressions of the Western Desert are associated with famous oases such as Kharga and Dakhla to the south, Farafra and Bahariya centrally, Fayum to the northeast, Qattara to the north and Siwa to the northwest. The southern ones are older than the northern and the scarps which determined their locations first appeared between the Cretaceous and the Eocene, forming after a retreat of the seas. Their excavation began in southern Egypt in the Early Oligocene.

Further development of the depressions may have involved running water and certainly chemical weathering which has been very active in Egypt since the Pliocene because of aridity. Rock solution and decay result from saline water percolating from sandstones in the swell areas, the capillarity of the saline solution producing accumulations of fine materials easily deflated by the wind. However, deflation alone cannot have excavated the depressions because the wind has only a limited mechanical ability as regards solid rock, although it can carry fine material. No doubt, deflation was essential as was the case in the Qattara Depression. This was excavated by the wind during the Pleistocene and Holocene. A northeasterly wind later pushed resultant sand and seif dunes southward into the Great Sand Sea.

Figure 12.2: The Ibis Temple in the Kharga Oasis, Western Desert.

Deepening of the depression stops at the water table and its level depends upon the quantity of groundwater and the sea-level (Figure 12.2). The elevation of the Mediterranean is related to the submarine outlet of the water which saturates the sandstones. It is believed that this elevation dropped 200 m during the Mindel Glaciation, sea level rising to -90 m during the later Würm cold phases. During the Mindel-Riss Interglacial, the Holstein, the level of the Mediterranean was higher than now by 35 m. During the Eem Interglacial between Riss and Würm, it was 15 m above the present level. This lowered the groundwater table and also the base level of the River Nile declined at least

120 m below the existing sea level. Foraminiferal and other faunal evidence from cores in the Red Sea may be correlated with radiocarbon dates. A climatic succession for the ice ages ends with the last Glacial. This lasted approximately 100,000 years and ended 16,000 years ago.
At Kufra, Landsat showed the existence of an ancient lake in the Late Pleistocene and Holocene of which four raised terraces remain between 400 and 430 m elevation. Associated artifacts testify to former humid climates. The earliest could have been the Kubbaniyan Pluvial and took place before 11,700 B.P. Afterwards, the Dishnian occurred between 10,500 and 7,100 B.P. Later, there were two more humid episodes, the first from 6,000 to 4,700 B.P. and the second, punctuated by arid phases, from 1,700 B.C. to 1,000 A.D. Thereafter, aridity set in and persisted until now (9).

Figure 12.3: A reconstruction of Sivatherium.

The Pleistocene lake at Kufra was located in a depression and was about 100 km long with a width of 60 km. It left deposits of rock salt intercalated with clay, silt and sand with some peat. Two salt lakes exist as remnants of it. The seasonal fluctuation of water level in these lakes and central sabkhas is alleged to be 33 cm. Most of the Kufra basin was heavily populated during prehistoric time which is confirmed by abundant stone implements and rock art which records stock breeding activities on a savanna as well as cultural pursuits. Also in the Pleistocene, the large giraffe Sivatherium, originating from Miocene African ancestors, roamed the savannas of southern Asia and Africa (Figure 12.3). It has been recorded from Tunisia, Libya and Djibouti. Curiously, it has not been described from Egypt. This animal became extinct at the close of the Epoch (10).

Rift lakes occur in Djibouti and Ethiopia occupying an enclave of internal drainage separating Blue and White Nile tributaries from the rivers flowing toward the Indian Ocean. The Ethiopian Highlands rising to 4,000 m and the Afar Depression are to the east of the Sahel and have a similar history of climatic fluctuation.

The Plio-Pleistocene in the Afar is marked by the development of extensive lakes and and the hominid-bearing Haddar Formation of west Central Afar in which "Lucy", a graceful australopithecine, was preserved. There are thirty beds containing an impressive collection of hominid bones as well as possibly the most ancient "tools" known. These latter are fractured pebbles, cores and flakes estimated to be 2.6 Ma old. Between then and 2 Ma B.P., aridity set in and desert conditions desiccated the Afar. This coincided with the development of major ice-sheets in the northern hemisphere and also with the first appearance of a flora adapted to summer drought in the Mediterranean. In fact, the climatic zonation of today appears to have been established in this period.

Extensive lakes with an archaic temperate flora of diatoms then occupied parts of the Afar and probably correlate with the Early Pleistocene lakes of the Hoggar. In the Afar Rift, the greater continuity of volcanism favored the preservation of areally restricted, varied deposits representing these lacustrine episodes. In the Middle Pleistocene, a period of major earth movements and continuing volcanism, Ethiopia and most of Africa had no lakes, the landscape probably evolving under arid conditions. Eolianites accumulated in the Afar Rift. In the Late Pleistocene, successive lacustrine highstands have been recorded from 40,000 B.P. until the lake levels fell rapidly after 20,000 B.P., Lake Abhé drying up completely by 17,000 B.P. This was much earlier than Lake Fayum in Egypt which desiccated 5,500 years later. Thus, a phase of pronounced aridity followed an episode of very high lake levels and covered the central part of the Afar to constitute one of the most distinctive climatic events in the region as well as in the Rift. The terminal Pleistocene aridity is sufficiently well established to serve as a possible basis for correlations across Ethiopia. At the southern end of the Afar, in one of the calderas of the Garibaldi complex, paleosols have yielded Middle Paleolithic artifacts as well as an ostrich egg shell radiocarbon-dated as 14,700 B.P., the local manifestation of the terminal Pleistocene arid phase (11).

During the Late Pleistogene, minor ice-caps and cirque glaciers developed on the highest Ethiopian elevations to leave moraines which indicate that the largest was on Mt Badda and covered 140 km². Core evidence provided a radiocarbon date of 11,500 years B.P. for the final retreat of the glaciers. This implies that the climate of the Early Holocene in Ethiopia was unfavorable for glacier readvance despite the return of more humid conditions because there was no substantial cooling. The ice-caps may have formed during the terminal Pleistocene arid phase and thus have corresponded with the maximal extent of the ice-sheets in the high latitudes. However, since lakes downstream from the Arussi Mountains are at very low level, precipitation at that time may have been too low to support ice-caps on the summits.

These climatic cycles and intricate tectonic movements must have influenced landscape and landslides which occasionally resulted in the Nile Valley or along escarpments. Oddly enough, large scale landsliding in arid North Africa has hardly been noticed. However, the phenomenon has been found on the north and west of the Murzuk Basin where huge sharp crested landslide segments often more than 1 km long and of rotational character are distinguished. This type of sliding is initiated by the underlying base of Triassic clays where these are thin. To the south, where they are thick and comprise most of the escarpment, a mudflow-type predominates. The landslides are relict features and developed after initial steepening of the escarpment probably at the turn of the Plio-Pleistocene. Humidity must have been sufficient to allow saturation of the clays. Final insignificant movements date from the Early Holocene.

Huge landslides are quite common in Upper Egypt, particularly between Esna and Qena as well as in the Valley of the Queens at Luxor. The rocks comprise Early Eocene limestones overlying Esna Shales under which occur Chalk and Dakhla Shales, the whole sequence reaching a thickness of 450 m. Usually several landslides, often three in number, followed each other. The movements started during the Early Pliocene, when the floor of the Nile Valley was 200 m deeper than it is today. However, most of the mass movements took place during the Early and Middle Pleistocene. The majority were from 100 to 500 m in width. Pluvials during this time contributed enough rain to saturate the Esna Shales which became thixotropic clays and created swelling pressures sufficient to buckle and fracture the overlying limestones and initiate landsliding.

Radiocarbon distribution in Saharan groundwaters demonstrates an alternating sequence of humid and arid episodes in the Late Pleistocene and Holocene. An observed west to east decrease of heavy stable isotope contents in Sahara waters is thought to represent the continental effect in both rain and groundwater. It is inferred that the paleoclimatology during groundwater accumulation was controlled by the western drift transporting wet Atlantic air masses across the Sahara. The best fit of a simple Rayleigh condensation model of the continental effect to the experimental data is obtained with a west to east decrease of the Saharan paleowinter precipitation of -30% per 1,000 km like the one in Europe today. With an annual winter precipitation rate from September to March of 600 mm for the North African coastal area at the Atlantic Ocean corresponding to Lisbon today, a previous winter rate of 250 mm is obtained for the Murzuk Basin (12).

In west Central Libya, there is fossil evidence demonstrating the past extent of a large Quaternary lake in which there were considerable salinity variations. Radiocarbon and uranium dating of shells show lacustrine periods at 160,000 to 125,000 B.P., 90,000 B.P. and 26,500 to 22,500 B.P. Fossil specimens of the euryhaline cockle Cardium glaucum, still living in the Mediterranean and bird-transported into Lake Qarun, have been recorded. The species has upper and lower limits of salinity of 0.3 to 0.6% and can also tolerate temperatures from 0 to 30 C as well as up to four times more calcium or potassium than normal seawater. Thus, it is not an indicator of a salinity close to that of normal seawater. A freshwater snail, Melania tuberculata, also tolerates high salt contents and is found in the Fayum today as well as fossil there and in other deposits of former lakes.

REFERENCES

1. MESSERLI, B., WINIGER, M. and ROGNON, P., 1980. The Saharan and East African uplands during the Quaternary. In: The Sahara and the Nile, ed. WILLIAMS, M.A.J., FAURE, H., 87-132, Balkema, Rotterdam, NL.

2. JUX, U., JUX, E., 1982. Diagenese quartärer Riffkarbonate aus dem Roten Meer.- Stockholm Contr. Geol., 37, 8, 99-115.

3. MOHAMED, M.A., Al-HAMLAN, A.A., 1979. The factors controlling the distribution of carbonate content in Kuwait Bay sediments. Geol. Rdsch., 68, 622-630.

4. BUTZER, K.W.,1980. Pleistocene history of the Nile Valley in Egypt and Lower Nubia. In: The Sahara and the Nile, ed. WILLIAMS, M.A.J., FAURE, H., 253-280, Balkema, Rotterdam, NL.

5. SAID, R.,1983. Proposed classification of the Quaternary of Egypt. J. Afr. Earth Sci., 1, 41-45.

6. CHURCHER, C.S., 1974. Relationships of the Late Pleistocene vertebrate fauna from Kom Ombo, Upper Egypt. Ann. Geol. Surv. Egypt, 4, 363-384.

7. BLUME, H.P., ALAILY, F., SMETTAN, U. and ZIELINSKI, J., 1984. Soil types and associations of southwest Egypt. Berliner Geowiss. Abh., A, 50, 293-302.

8. PACHUR, H.-J., RÖPER, H.-P., 1984. The Libyan (Western) Desert and northern Sudan during the Late Pleistocene and Holocene. Berliner geowiss. Abh., A, 50, 249-284.

9. EL RAMLY, I.M., 1980. Al Kufrah Pleistocene lake - its evolution and role in present-day land reclamation. In: The Geology of Libya, 2, ed. SALEM, M.J., BUSREWIL, M.T., 659-670, Acad. Press, London, UK.

10. GERAADS, D., 1985. Sivatherium maurusium (POMEL) (Giraffidae, Mammalia) du Pléistocène de la République de Djibouti. Palaeont. Z., 59, 311-321.

11. GASSE, F., ROGNON, P. and STREET, F.A., 1980. Quaternary history of the Afar and Ethiopian Rift Lakes. In: The Sahara and the Nile, ed. WILLIAMS, M.A.J., FAURE, H., 361-400, Balkema, Rotterdam, NL.

12. SONNTAG, C., KLITZSCH, E., EL SHAZLY, E.M., KALINKE, C. and MÜNNICH, K.O., 1978. Palaeoklimatische Information im Isotopengehalt ^{14}C-datierter Saharawässer: Kontinentaleffekt in D und ^{18}O. Geol. Rdsch., 67, 413-423.

Chapter V-13

ENIGMATIC SAHARA AND OTHER DESERTS

"And the hills are ground to powder so that they become a scattered dust".
Surah of the Event, 56.

A CHIAROSCURO OF ARIDITY

Probably the Sahara operates as a dynamic unit as regards wind action, sand movement harmonizing with trade-wind circulation except where major topographic obstacles cause sand currents to move against the regional wind direction. It is important to recognize this latter phenomenon and not misinterpret it as representing an earlier wind regime (1).

Eolian action produces characteristic diverse landforms according to whether sand is transported or deposited. In the first case, deflation zones with barchans, wind abraded ridge-and-swale systems and yardangs together with serirs and hamadas are diagnostic. As for the second, in sand seas or ergs specific types of dunes reflect the interaction of particular sets of eolian processes. From Central Sahara to the Sahel, there is a general decrease in sand particle size which may account for the successful cultivation of cereal grasses in the latter (2).

Precipitation in the Central Saharan lowlands is less than 50 to 100 mm annually. This is perhaps the driest place on Earth with an aridity index of 200 which means that the incident solar energy can evaporate 200 times the amount of precipitation received. Hence, it is quite understandable that the evaporation potential at Kharga Oasis is 5,000 mm per year. Though small, the rainfall nevertheless must have an important influence on the geomorphology. This reflects the rainfall decline from west to east manifested also in the continental effect shown by the heavy stable isotopes. There is an increase with elevation also which is demonstrated by the altitude effect on deuterium and oxygen-18. Therefore,the Hoggar and Tibesti Massifs receive a yearly precipitation comparable with that of the Sinai Mountains, i.e. 100 to 200 mm (3).

Stable carbon isotope ratios of organic matter in rock varnishes of Holocene age from Israel and Sinai show a strong relationship to the environment. This organic material occurs as a minor component of rock varnishes primarily made up of Mn and Fe oxides, clay minerals and trace elements. Such varnishes in deserts are biogeochemically stable with slow formation rates which facilitate their use in dating and recording alkalinity fluctuations. The carbon-13 isotopic variability reflects the abundances of plants with different photosynthetic pathways in adjacent vegetation. Analyses of different layers of varnish on Late Pleistocene desert landforms show that the carbon isotope composition of their organic matter is a paleoenvironmental indicator. Carbon-13 values from arid sites range from -12 to -19‰, those from more humid localities having values around -23‰. Desert varnishes from arid sites have carbon-13 isotope ratios reflecting the presence of so-called C-4 or CAM plants or both in adjacent vegetation (4).

Curiously, Mount Hermon in Syria, the traditional site of the Transfiguration of Jesus, has a yearly average of more than 1,500 mm although it is only a few hundred kilometers north of Sinai. Even though not protected by any other orographic barrier against advancing cyclones, the location of Mount Sinai in relation to the Mediterranean coastline and the path of rain-bearing ectropical depressions is responsible for its arid state. Rainfall is mostly in the winter, responding to the passage of westerly depressions reaching the Atlas Mountains from the Atlantic and weakening rapidly inland. Such depressions sometimes reach the Hoggar or even the Tibesti to produce light but long-lasting rains. Heavier rainfall takes place when colder ectropical air masses meet warmer humid equatorial air flows. Summer rainfall in the northern Sahara almost always takes place during thunderstorms and hardly ever occurs in the Sinai.

Certainly the most tedious of the geomorphologic features of the Sahara are its great sand sheets which cover large stretches, such as in the southwest of Egypt where they form flat, featureless areas. They comprise medium to coarse-grained sand veneered by small pebbles or coarser material and possess hard compact surfaces. Their thickness is not well known, but at Bir Sahara reaches 22 m for loose sand overlying sandstones. Certainly, it must vary considerably in the region. All the sand sheets have a reddish-brown soil which, to judge from artifacts found at the above oasis, is pre-Acheulian in age. Dunes have been forming on the sand sheets since the Neolithic and today are often parallel and separated by flat deflational streets.

The average size of sand grains in the coarse fraction of these sheets is from -1.0 to -2.5 phi. For comparison, the average size of desert lag varies from 1.4 to -1.0 phi. As regards their origin, it is most likely that the sands derive from erosion of highlands and transportation through several fluvial cycles to be deposited in coalescing alluvial fans, cones or terraces. These finally compose the great trunk channel extending from the Gilf Kebir to Siwa Oasis and beyond. Modern dunes lying along it. Evidence for alluvium is provided by coarse pebbles strewn over the sand sheets today, especially along the Egyptian-Libyan border at about the 27th parallel. This is the site of the Great Sand Sea where the Libyan Desert Silica Glass occurs. After these events of the Dishnian Pluvial, deflation bevelled the sand sheets which later stabilized. Subsequently, the dune accumulation started and is still going on.

Freshwater fossils have been recorded from the sand sheets south of Siwa Oasis in Egypt and may be of Neogene age. In the center of the Sirte Basin of Libya, a Middle Miocene mammalian fauna has been recognised. Additionally, this region is rich in Roman dams constructed after 69 A.D., most of which were first noticed when Italy colonized the country. In 1929, a detailed survey of them was carried out as a preliminary to a Fascist revival of the Roman Empire. These ancient barrages are variable in type, often comprising a rubble and gravel core bonded with lime mortar and faced with limestone blocks or cobbles and plastered. Sediments trapped by them have given clues to the nature of stream flow in Roman times. However, Tripolitania has nothing to compare with the many Himyaritic inscriptions mentioning irrigation in southern Arabia from 700 B.C. up to the present. In the latter part of the 1st century A.D., Sextus Julius Frontinus referred to the building of dams as an African tradition. Not all were built to conserve soil and water, some being made to keep seawater out. They all rest on calcrete foundations. In Cyrenaica, stone dikes were built in classical antiquity to curb spates.

Some parts of the sand sheets may be blanketed with a continuous carbonate layer which displays large polygonal cracks, probably a remnant of leaching action and formerly much more extensive. The largest occurrence is from Bir Sahara to Bir Tarfawi, covering as much as 4,000 km². Such carbonates were precipitated during the Pleistogene pluvials, the oldest during the Acheulian culture when springs flowed. The connotation is that the area was then a basin with a high water table. Subsequently, this declined and calcification of organic matter took place. Later, the water table rose again and, during Mousterian and Aterian times, the basin refilled, lacustrine deposits accumulated and vegetation burgeoned and luxuriated. Probably by a similar mechanism, but under different chemical conditions, silcrete pavements grew (5).

Marlstones, peat deposits, tufas and salt encrustations represent spring deposits and often record several phases of activity. They occur near present-day oases or wells, but must have existed elsewhere and their deposits have been eroded away. The oldest are those on the top of the limestone plateau in the Kharga, Kurkur and Dungul Oases. They are up to 20 m thick and comprise crystalline carbonate rock containing molds of plants. Their age is probably pre-Pleistocene. Younger ones found elsewhere are perhaps Middle Paleolithic in age since they contain Mousterian implements.

Former spring activity is revealed by shallow round depressions and vents. The circular features vary in diameter from 15 to 100 km with a maximum depth of 40 m. They may have been caused by settlement of water-bearing sandstones after the disappearance of older springs. In some of the lows, deflated playa deposits with Mousterian artifacts show that many are earlier than this in age.

The spring vents comprise mostly conical mounds which may rise to 20 m in height and are cylindrical in shape. The restraints on their formation probably parallel those of similar modern features found at the foot slopes of the escarpments of the limestone plateau. Some Roman wells lie away from these and are 10 m higher than modern wells because of the later lowering of the water table. Some thermal springs emit ferruginous waters. Other silica-rich ones produce plugs of quartzite which fill in their vents or fissures at the last stages of activity.

Examples of this are in the Wadi Natrun, where the waters gve an oxygen-18 value of -7.6‰ near to that of the Nile at -8.8‰, and, more conspicuously, at Gebel Ahmar, the Red Hill in Cairo. Pharaonic quartzite quarries hewn out of this hill provided material for royal sarcophagi and valuable ornaments. Strange-shaped cylindrical structures were eroded here from the Oligocene sandstones and display the results of upwelling fluids emerging first as fumaroles and then as hot mineral springs. However, most of the desert wells are fresh-water. Their fossil equivalents are easily recognized, but some have lost their original shapes and are found as tabular hills of dark sandstone resembling buttes. The contradictory term "inverted wadi" was introduced to describe elongate, serpentine and occasionally branching gravel ridges which some consider to be old thalwegs. They are all found around the Nubian Nile and mostly exceed 10 km in length. Rolled Acheulian artifacts have been recorded from some of them.

Dry lake beds, commonly known as playas or vloers, constitute the remnants of Holocene pluvial water bodies. At least fifty sheets of them lie mostly at the foot of the great Libyan Limestone Plateau and represent the lowest points of the various enclosed drainage basins of the desert. All have scanty vegetation and are composed of fine-grained sediments. Such clay silt features are called hattiya in Egypt, this meaning "temporary settlement". Any water falling over the desert accumulates in them, so that roving nomads can stay for a while. Usually, they are uneconomic except for the region around Selima in the Sudan where nitrate and alum (hydrated potassium aluminum sulfate) deposits have been recorded. An individual hattiya might cover several hundred square kilometers and all of them have deep water tables. Most of them originated during the Dishnian Pluvial and were not related to springs. Nevertheless, a few lakes connected with springs persisted into Roman times, such as the Kharga Oasis lakes where antique pottery has been collected from the youngest lacustrine deposits. The associated vloers are unique in that they were still forming so late in history. They cover most of the depression, extending almost 90 km and making the New Valley Resettlement Project feasible. It appears that the water table here is declining at about the same rate as the lowering of the desert floor, i.e. of the order of 4 m per millenium. The deflation rate is measured by the height of the yardangs of playa deposits left behind. Some hattiyas, termed armored playas, are canopied by white nodular chalcedony cobbles. Since some include Acheulian artifacts, they must be Late Pleistocene or younger in age.

The best-known eolian landforms in the depression floor of the desert are the sand dunes which, in northeastern Africa, align with prevailing winds. In the southern Sahara, the main dune-forming winds are the NE Trades, the modern Harmattan. In the eastern Sahara, the sands constitute the bed load of an enormous braided air stream which branches around obstacles such as the Gilf Kebir and Gebel Uweinat. Various types of dune exist such as the whalebacks of the Great Sand Sea which separate southward into parallel longitudinal seif dunes. These take their name from the Arabic word for "sword". They are sharp-crested with curved slip faces on one side, these tending to increase their height and width which may range up to 200 m and 100 km respectively in Egypt. As well as these and often aligned along the same arcuate stream lines, discontinuous elongate clusters of crescentic dunes called barchans of Late Holocene to modern age occur. Their age is determinable because they override Middle and Early Holocene sand sheets. Actually, three other major dune generations have been recognized from the Late Pleistocene and are thought to be 100,000, 40,000 and 20,000 years old. Naturally, the oldest are the least well preserved, but so long as they can still be identified, it is apparent that they were produced in large and well-ordered fields (6).

Barchans are named from the Turkish word for sandy hills. The depressions between them are termed "fuljis" (sing. "fulji" or "fulje") in northern Arabia. These solitary dunes can move up to 100 m in a year. Consequently, they frequently interfere with the activities of Man. In such cases, they ought to be stabilized in their uppermost 20 cm of loose sand. Various binding agents can be used such as vegetation. Although they are often deficient in nutrients, dunes can be preferentially colonized by plants because their high porosity and permeability allow appreciable water storage. Of course, any precipitation falling upon fixed dunes is absorbed into them, surface run-off being extremely rare. Such crescentic dunes are orientated transversely to the direction of the prevailing wind and have a gently inclined convex side facing it. Hence, the horns point downwind and there is a steeply sloping concave or leeward side inside them. Heights of more than 30 m and widths up to 350 m can be attained. In the Kharga Depression, these horseshoe-shaped accumulations of sand are arranged into three belts paralleling the longitudinal axis. Forms vary according to their maturity. Within an area of 27 km², the total number of dunes counted was 261, but widely varying numbers have been found elsewhere in similar areas. Dune densities increase downwind and merging may occur while individuals increase in size. Most of the sand grains in the barchans are between 1.3 and 3.3 phi in size, but coarser grains are found on the foot of the windward side due to sorting processes occurring there and at the crest. The roundness values increase with the grain sizes. Similar characteristics have been found in dunes throughout the world. Other types of sand dunes include the immobile star, the mobile silk and various non-barchanoid transverse ones such as the akle, parabolic and mreyye which sometimes form very extensive fields.

"Slides the silent meteor on, and leaves a shining furrow" said Tennyson, which vividly describes the brilliant night sky so familiar to every traveller in the desert. Thunderstorms and their effects are more rarely seen. Nevertheless, like meteorites, they leave behind their fascinating traces as well. These are lightning tubes or fulgurites which are frequent in dune areas of deserts like the Great Sand Sea. They comprise vitreous, irregular, rod-like and fragile structures or crusts. During sand-storms, lightning strikes and fuses loose sand together by impact to shape it into tubes sometimes several meters long and 6 cm or more in diameter. This reflects the transmission of the charges through dunes to the humid ground surface often ending in flat plates at the bases of the glassy sand tubes.

Coastal dunes are found along the North African shores, but differ in that they are composed of calcareous sand rather than quartz grains like the inland ones. Also, they are orientated east-west and have a seif-like appearance. Their mean grain size is 2.1 phi, hence smaller than the average for the inland dunes.

THE SPELL OF THE WESTERN DESERT

Six weeks sufficed to drive the slave caravans through the southern part of the Western Desert from

Sudan to markets in Egypt, where those in bondage were sold to the highest bidders as late as the turn of the century. Today, this poignant trail followed by the slave masters and their pitiful prey is called the Darb el Arbain, the Road of the Forty Days. That part of the Western Desert through which it runs constitutes the Arbain Desert on the Sudano-Egyptian border. It is a landscape molded by wind and devoid of integrated drainage systems.

There is evidence of past periods of wetter climate separating longer intervals of hyperaridity. Alternations between humidity and desiccation together with changes in differential erosion were major factors in shaping the scenery.

Deflation led to oases depressions which reached their present outlines and depths before 200,000 years ago when the Late Acheulian people used their artesian springs. Two major pluvials have been recognized, the Kubbaniyan during Mousterian-Aterian times and the Dishnian in the Holocene from 9,500 to 6,000 B.P. The Mousterian reached its acme 50,000 to 30,000 years ago when springs, lakes and grasslands covered much of the Sahara. This flowering of Middle Paleolithic cultures left relics all over Egypt. Arkin 5 at Kharga has yielded a vast number of artifacts and such accumulations are common in shallow sites where hunters must have returned from time to time, pitching tents and making and sharpening tools. It is peculiar that no such accumulation occurs at the Mousterian site in the Bir Sahara Depression. This may be because there was a different social pattern there from those in other Middle Paleolithic settlements. The succeeding Aterian culture developed arrowheads and spear blades nearly 22 cm long. Thus arose a Paleolithic group superior to its contemporaries and capable of improving on their sling-stones by using the more accurate and deadlier bow and arrow. In the Khormosan culture, which ended the Middle Paleolithic, many sharp burins or arrowheads are found. They were used to tip arrows or darts in the Egyptian Late Paleolithic and probably performed the same function in the earlier Khormosan industry.

Some terminal Paleolithic and Neolithic sites have been discovered at Nabta Playa, 100 km west of Abu Simbel. They have been dated between 7,300 and 4,000 B.C. Fossil dunes, heavy clays and silicified root casts found today witness to the effects of a subpluvial of the Dishnian in the Neolithic. This was of great importance as its rains made extensive human occupation feasible in the Sahara for the first time in 30,000 years. It is probable that the latest Paleolithic people moved into the area around 9,000 years ago and remained there until at least 6,600 B.C. hunting gazelle and hare. However, this way of life was soon replaced by a Neolithic farming and herding economy. It was founded on the cultivation of barley and the domestication of sheep, goats and cattle. Despite fluctuations in the rainfall, the settlement persisted in isolated spots like Nabta for at least 2,000 years.

The Neolithic life-style began in the northwest of Africa under cold wet climatic conditions from the 5th to the 3rd millenia B.C. when caves occupied in summer by sheep and goat herders were abandoned in the winter. Even 2,000 years earlier than this, wetter conditions had prevailed from the Atlantic to the Nile until the Sahara commenced to dry up. This ominous paleoclimatic event drove these herders into the mountains, along lake shores and on to the banks of the Nile. Neolithic communities in the Central Sahara became confined to isolated oases. Famines must have occurred often, the earliest written record being that of the bas-relief of the Fifth Dynasty Pharaoh Unas. However, conditions stabilized later and remained fairly constant during the past 2,000 years. The "Little Ice Age" lasted from the 16th to 18th Centuries in Europe and probably coincided with a glacial advance. During it, the climate of Africa was analogous to that of the Neolithic. Increased rainfall took place along the tropical margin of the Sahara, in the desert and possibly in North Africa as well. This could have helped powerful realms such as the Hausa States of northern Nigeria or the Songhay Empire of northern Mali to arise (7).

Severe droughts occurred in Africa during the more arid conditions of the 18th century A.D., the first of which in the 1680s was called the "seven year" famine and affected the Sahel and Sudan. More humid conditions returned in the late 19th Century . Thus, at Aswan, the annual discharge of the Nile was on average 109 billion tons between 1870 and 1899. In the 20th Century, this dropped to only 83 billion tons from 1900 to 1949 showing a return to drier conditions. Only from late Predynastic times up until today can human beings have influenced marginal desertification in the Sahara. Paleolithic Man exerted hardly any effect on the world around him. Even during the Neolithic, important landscape and ecologic changes arose mostly through natural climatic variations (8).

There is evidence for Neolithic animal life in the Gilf Kebir and in Gebel Uweinat as shown by faunal remains of elephant, ox, gazelle, goat, jackal, ostrich, wild donkey and dog and also by rock drawings. Of course, the provenance of the remains is unknown, but probably some were associated with cultural materials which have been identified at Wadi El Bakht and some other places. At this wadi, an ostrich egg shell gave a radiocarbon date of 7,280 B.P. Rock engravings at Gebel Uweinat identify five phases of former occupation. In order of age, these are wild herbivores and ostriches (oldest), game and domestic cattle, cattle (mostly long-horn), short-horn cattle and later goats and dromedaries and other domesticates.

The present mammalian faunal remains at Gebel Uweinat and Gilf Kebir are limited to Barbary sheep and gazelle. In 1969 a Toubous family with goats, dromedaries and some semi-wild donkeys was living at Uweinat. Half a century ago, the area was regularly visited by small numbers of such shepherds. Degradation of the environment started through climatic change, but probably continues because of the destructive effect of Man and his livestock.

Figure 13.1: Neolithic Man and his dogs hunting the ibex in Wadi Abu Pissale of the Eastern Desert.

From the Abu Hussein dunal field, bones of domestic cattle have been collected, but their origin is obscure.

The first explorer of the Gebel Uweinat, Ahmed Mohammed Hassanin Bey in 1923, recorded ancient rock art demonstrating that the climatic regimen when it was created was much moister than now. As the camel is not found in any of these drawings, he concluded that they dated from before the introduction of that beast of burden to the region around 500 B.C. In fact, they are much older. Pastoral motifs in the Ennedi and Tibesti Massifs of eastern Chad were carved and painted earlier than 5,000 B.C. Possibly such art accompanied a slow transition from hunting to cattle rearing (Figure 13.1). This may mean that domestication did not just diffuse in from the Middle East, but developed independently in some parts of North Africa. Most likely, as among the Nuer of the Sudan today, the cattle were herded by young unmarried youths, not women.

Radiocarbon dates from many North African sites suggest that big game hunting was gradually replaced by pastoralism when the Dishnian Pluvial reopened this part of Egypt to widespread human settlements. Nomadism in the Arbain Desert is coupled with pluvials which may well be out of phase with the Late Pleistocene glacial stages of the Northern Hemisphere. The Pleistocene arid intervals seem to have been dryer than now and were characterized by extreme deflation. Remnants of relict terra rossa soils in solution cavities on the Egyptian Plateau intimate the presence of an Early or pre-Pleistocene karst terrain.

During the Late Cretaceous and Tertiary, peneplanation occurred in the Kufra Oasis area. Eroded sands and gravels, a few meters thick, became lateritic during warm, moist climatic conditions and hardened into an ironcrust during a semi-arid climate. On the slightly higher erosional planes, i.e. those without laterite, silicrust developed. Later, when more humid conditions returned, the erosion of the silicrust, the ironcrust and the Mesozoic rocks was greater because of a possible slight epeirogenic uplift. As a result, breccias and fanglomeratic sediments were deposited and later lateritized and hardened into a second ironcrust. Erosion continued and the peneplane diversified into mesas and inselbergs. During the Early Quaternary, the erosion was interrupted and coarse-grained material deposited. This later lateritized into a third ironcrust.

The fossil river system is extensively branched and drained the whole of Kufra from southwest to northeast. Much of the eroded material flushed away and only some very thin fluvial beds remain. After the climate changed from humid to arid, intermittent fluvial erosional activity ended. Fine clastic limnic sediments deposited in local depressions and often alternate with eolian sands. In this transition period alluvial fans accumulated at the foot of mesas. An arid desert climate probably began about 4,000 B.C. (9).

THE EMPTY QUARTER

Known as the Rub al Khali,this is a vast body of sand blanketing almost 600,000 km² of Saudi Arabia and, together with the 180,000 km² of the interior nafuds and Ad Dahna, covering approximately half of the country.
The various sand terrains include transverse, longitudinal, uruq or seif and sand mountains. The transverse type is composed of predominantly simple and accreting barchans in areas of more mobile sand sometimes with rounded ridges lying transversely to the prevailing wind direction. The longitudinal ones are mainly of dikakah type with bush- or grass-covered sand. In addition, there are undulating sand sheets elongated in the direction of the prevailing wind and often stabilized by scanty vegetation. Uruqs comprises long, subparallel sharp-crested narrow sand ridges together with dune chains separated by broad sand valleys which may include transverse features as well. These forms result from two dominant wind directions. Sand mountains may rise up to 300 m, often with superimposed dune patterns built by complex barchans. The commonest ones are enormous and span several kilometers from horn to horn. Also common are giant sigmoidal and pyramidal sand peaks as well as large oval to elongate sand mounds.
Terraced lake beds exist in the southwest of the Empty Quarter. The freshwater sediments contain fragmentary vertebrate teeth and bones of maybe Late Pleistocene to Early Recent age. Elsewhere, Paleolithic and Neolithic sites have yielded arrowheads, hatchets, scrapers and other stone implements. At one locality, charcoal from a presumed Neolithic campsite gave a radicarbon age of 5,000 years. The Rub al Khali Basin is mainly a Tertiary feature with

sedimentary rocks thickening moderately toward its center, but in East Oman much greater deposition occured in a deep trough in front of the intensely folded Oman Ranges. The underlying Mesozoic rocks are shallow marine toward the shield, but to the northeast deeper water conditions prevailed. It is interesting that the Mesozoic sedimentation involved a much greater basin than the present depression. If so, the Rub al Khali as an independent basin is exclusively Tertiary.

From Arabia into Iraq and Kuwait, there is a vast area of sheet gravel representing the residue of a tremendous flood of rock debris originating from the basement complex and funnelling out through a wadi channel system. This is nowhere more than a single pebble thick and is mostly composed of quartz and basalt.

THE SALTS OF THE EARTH

Sabkha or sebkha, plural: sibakh, is the Arabic word for coastal and inland saline flats or playas accumulated by the deposition of alluvium in

Figure 13.2: Nomad hacking out halite from a sabkha in the Selima Oasis, Sudan.

shallow, sometimes extensive, depressions. The deposits are usually saturated with brine and are often salt encrusted. Actually, salt is the characteristic of a sabkha. Those foci of centripetal drainage usually containing silt, but no salt, may be distinguished as qian. A sabkha where salt is being or was mined is referred to as a mamlahah, plural: mamalih, this, strictly speaking, applying to the relevant place of excavation (Figure 13.2).

Sibakh are concentrated mostly in a narrow belt along the Persian Gulf Coast from Kuwait to Qatar, the majority being found within 60 km of the shoreline. However, some occur far inland and there is a wide band along the eastern periphery of the Rub al Khali. They may contain anhydrite nodules which resemble calcium sulfate nodules in ancient rocks. This indicates that the latter could have had the same origin (10).

Considerable quantities of wind transported material occur in some Persian Gulf deposits. The importance of this contribution has been underestimated in the past, but fallout from individual dust storms shows that probably more than 2 cm per year is involved. Much of this goes to the northern part of the Gulf and correlates with carbonate content and grain size interrelationship. Gypsum particles arise from the entire region of flood plain soil in the Mesopotamian Plain.

There is an offshore prograding sand sea along parts of the Gulf coast in Dhahran. Here, eolian and siliciclastic sabkha sediments intercalate with marine deposits. Such interfingering creates optimum conditions for local petroleum traps caused by a facies change from porous, permeable dune sands to impermeable, marine mudstones. The setting is controlled by strong offshore winds supplying abundant sand to the coast line and now causing the construction of the dune system. The eolian system itself includes potential reservoir rocks, perhaps sabkha sources and seals comprising zones of early cementation in the deposits which are very common in all facies of the eolian sand sea. They occur as a result of soil formation, deposition of pore-filling gypsiferous cements from saturated solutions near the water table and the addition of sand-sized, windblown, evaporitic material to sands downwind of sabkhas.

The origin of the sibakh is not fully understood, but brine always occurs within a zone 1 or 2 m below the surface. It seems that prolonged evaporation of capillary water concentrates and precipitates various salts in this zone to produce a hard saline crust. In moist periods, stagnant water from tidal invasion or run-off aids in recharging sediments and, partly through its evaporation, adds to the salt scum.

Dolomitization of aragonitic intertidal sediments in the subsurface supratidal environment of a sabkha along the south shore of the Persian Gulf near Abu Dhabi is proceeding as a result of percolation of brines. Abundant diagenetic dolomite is developed from a combination of factors such as high Mg:Ca ratio fluids, rapid flow rate and optimum shoreline configuration. Dolomitization lasted up to 1,500 years. The sabkha is prograding and has engulfed some islands which are strongly dolomitized on their leeward sides (11).

Large tracts of northwestern Arabia are masked by a resistent calcareous duricrust. How this formed is not quite clear, but it may have been a product of a rather moister earlier climate. During the dry season, groundwater rises to the surface and deposits soluble substances. During evaporation, cal-

Figure 13.3: Honeycomb weathering of cross-bedded Carboniferous sandstone in Wadi Hawashia, Sinai.

crete builds up and may be so resistant as to support 50 m high escarpments. Where sandstones rather than carbonates are involved, the result is much less durable.

Duricrust is especially well developed on Paleozoic and Mesozoic units in the Jawf and Widyan Basins as well as on Tertiary beds. It is usually a discontinuous mantle of varicolored, resistent sandy limestone containing fragments of underlying rock. The thickness of the crust ranges normally between a few centimeters and 3 to 4 m. In the southwestern Rub al Khali, the few minor duricrust patches are gray or tan in color and comprise locally gypsiferous, sandy limestone grading sometimes into sandstone with calcareous cement. Duricrust is not known to have formed in rocks younger than the Miocene and Pliocene, so these carbonate-enriched rocks are considered to be Quaternary (Figure 13.3).

In northeast Africa, similar sabkha-like bodies, but with permanent saline water in them, are found. Some of the best known of these are in Wadi Natrun in Egypt. There, the water may influx by seepage from the Nile and natron forms by evaporation. This is a highly soluble decahydrate of sodium carbonate which also precipitates in northern Sudan, Kenya and the western USA. The resource is exploited today as it has been ever since Pharaonic times when it was used for manufacturing glass and mummification. Its name derived from the old Egyptian word "ntrj" and the wadi itself is said to have contained a secret mountain of Osiris in Ptolemaic times. Earlier during the Middle Kingdom, Wadi Natrun was a commercial center not only for natron, but also for halite, skins of leopard and wolf as well as medicinal herbs. In the Christian era, over forty monasteries were established and many were destroyed by the Bedouins.

Other salts deposited by evaporation of sea and lake waters in saline playas include rock salt and gypsum in the Nile Delta together with alum and magnesium sulfate in the Dakhla and Kharga Oases as thin bands in black shales of the Nubian sandstones.

REFERENCES

1. MAINGUET, M., CANON, L. and CHEMIN, M.C., 1980. Eolian landforms of the Sahara. In: The Sahara and the Nile, ed. WILLIAMS, M.A.J., FAURE, H., 17-35, Balkema, Rotterdam.

2. SAID, R., 1983. Remarks on the origin of the landscape of the Eastern Sahara. J. Afri. Earth Sci., 1, 153-158.

3. MESSERLI, B., WINIGER, M. and ROGNON, P., 1980. The Saharan and East African uplands during the Quaternary. In: The Sahara and the Nile, ed. WILLIAMS, M.A.J., FAURE, H., 87-132, Balkema, Rotterdam.

4. DORN, R.I., DeNIRO, M.J., 1984. Stable carbon isotope ratios of rock varnish organic matter: A new paleoenvironmental indicator. Science, 227, 1472-1474.

5. HAYNES, C.V., 1982. The Darb El-Arbain Desert: A product of Quaternary climatic change. In: Desert Landforms of Southwest Egypt: A basis for comparison with Mars. Ed. EL-BAZ, F., MAXWELL, T.A., NASA, CR-3611, 91-117.

6. FRYBERGER, S.G., AL-SARI, A.M. and CLISHAM, T.J., 1983. Eolian dune, interdune sand sheet, and siliciclastic sabkha sediments of an off-shore prograding sand sea, Dhahran area, Saudi Arabia. AAPG, Bull., 67, 280-312.

7. GAUTIER, A., 1982, Neolithic faunal remains in the Gilf Kebir and the Abu Hussein dunefield, Western Desert, Egypt. In: Desert landforms of southwest Egypt: A basis for comparison with Mars. Ed. EL-BAZ, F., MAXWELL, T.A., NASA CR 3611, 335-347.

8. GABRIEL, B., 1980. Desertifikation der Sahara in der Vorzeit? Geomethodica, 5, 81-108.

9. BECKER, R.E., 1979. Die tertiäre und quartäre Entwicklung im Bereich der Kufrah-Oasen (Zentrale Sahara). Geol. Rdsch., 68, 584-621.

10. WEST, I.M., ALI, Y. A. and HILMY, M.E., 1979. Primary gypsum nodules in a modern sabkha on the Mediterranean coast of Egypt. Geology, 7, 354-358.

11. PATTERSON, R.J., KINSMAN, D.J.J., 1982. Formation of diagenetic dolomite in coastal sabkha along Arabian (Persian) Gulf. AAPG, Bull.,66, 28-43.

Chapter V-14

WATER RESOURCES AND RADAR RIVERS

"Thou preparest the waters under the Earth and Thou bringest them forth at Thy pleasure to sustain the people of Egypt even as Thou hast made them live for Thee". Pharaoh Akhenaten (14th Century B.C.), Hymn to the Aten.

A LAND OF DEATH

In Pharaonic times, the Western Desert was known under this name which may be applied also with some felicity to arid regions in general. Of these, the Sahara and the Arabian deserts rank among the greatest. Located mostly between latitudes 10° N and 30° N, they cover 10 million km² and are dominated by siccative climatic conditions. Despite this, more than 100 million people live in them. The countries concerned are almost totally dependent on groundwater for domestic, agricultural and industrial requirements. Despite the desiccation of the landscape, in the eleven major sedimentary basins, eight in North Africa and three in Arabia, there may well be at least 80 billion km³ of groundwater. There are common geological features throughout the region. The Precambrian is usually an igneous-metamorphic basement on which sediments were deposited, generally in huge intracontinental or paralic basins, and were intercalated by various thicknesses, extensions and ages of Phanerozoic volcanics. The Alpine Orogeny affected the region and the central part is split in a NW-SE direction by the major East African Rift Valley, partly occupied by the Red Sea.

The desert area has an arid zone hydrology, the interesting beds being the sandstone facies of the Paleozoic and Mesozoic known in the east as the Nubian and in the west as the Continental Intercalaire. Thicknesses range over more than 3,000 m. In the Arabian Peninsula, the sandstone thickness averages 1,250 m with the water under both unconfined and artesian conditions in different places and TDS under 4,000 ppm in the west and 21,000 ppm further down the hydraulic gradient to the northeast. In regard to this, the isopiezometric surface is inclined, thus indicating recharge in regions where it is elevated and discharge at localities where it is lowest. In northeast Africa as well, the beds are often confined and occasionally artesian. Radiocarbon dating yields ages in the 40,000 to 20,000 year range, but one He-Ar date gave 850,000 years. Probably, the recharge of the aquifer is now appreciably slower than during the pluvials and earlier geological history.

In North Africa and in Arabia, the sandstones are overlain by fissured limestones and dolomites which function as another aquifer, recharging by the upward seepage from the underlying sandstone aquifer. These carbonates attain a thickness of 4,000 m and belong to the Mesozoic and Tertiary (1).

As regards the North African sandstones, up to 10 N latitude, the beds are almost horizontal and practically undisturbed with the basement usually not deeper than 300 to 500 m. North of 20 N latitude, large-scale tectonism took place with uplifts, these being followed by downthrusts into deep sedimentary basins. The aquifer depths in these latter can reach 2,000 m and the total aquifer thickness often exceeds 500 m. Although radiocarbon ages range between 40,000 and 20,000 years as stated above, ages from 10,000 to 7,000 years have been recorded in some places. Several mechanisms for maintaining the flow of fossil aquifers under conditions of zero recharge have been described. The fossil groundwater has been exploited intensively in the region since the Second World War with a result that water levels have fallen and, in Egypt, at least half of the more than 700 wells in the New Valley project area stopped flowing and have had to be pumped (2).

Radiocarbon ages of groundwater from oases in the New Valley have been determined as exceeding 20,000 years. However, from the basement high between Gebel Uweinat and Aswan, younger groundwater is probably from the post-Dishnian Pluvial because it has been dated as less than 14,000 years old. Deuterium and oxygen-18 contents are said to indicate that the groundwaters from this region come from the same source as those of the New Valley oases but this is highly improbable. The groundwater is richer in the heavy isotopes because of evaporation which is thought to act at deeper levels than earlier

assumed. It is noteworthy that the stable isotope content of Saharan groundwaters is very similar to that of recent European ones, the deuterium excess being +10‰ in both cases and the usual meteoric water line relationship being followed (3,4).
Data available for solving arid zone hydrologic problems are far from adequate and many are highly unreliable because of the relatively small number of measurements taken from aquifers and both the infrequency and short duration of pumping tests. Difficulties stem from human impact. For instance, wells such as those at Kufra in Libya together with Kharga and Dakhla in Egypt as well as in Saudi Arabia have suffered uncontrolled drilling and pumping. In addition, brackish water or seawater may be injected into oilfields or water sources.

NORTHEAST AFRICAN GROUNDWATERS

Rainfall in Libya over the last two decades reached as much as 393 mm annually in the Benghazi Plain, where there is an evaporation potential of over 1,700 mm on grass-covered areas. The average yearly rainfall surplus, comprising run-off and infiltration, at Benghazi is thought to be 18 mm. Most of it takes place through karstic channels. An optimal area for studying this is at the largest spring in Libya, the karstic one of Ain az Zayanah on the northern part of the Benghazi Plain. This is the outlet of an aquifer in the Tertiary limestone and discharges 5.5 m^3/sec of brackish water. There are more than five independent groundwater systems in Libya. Irrigated agriculture developed rapidly in recent years so that by the end of the century the extraction will probably exceed 2 billion m^3 per year (5).
There are two main sedimentary basins in eastern Libya which are the Sirte and the Kufra. The top horizons in each, post-Eocene in the former and Cretaceous Nubian in the latter, are regional hydrogeological systems. The Sirte basin deposits include mostly fluviatile sands and clays in the south with marine, in part sandy, carbonates together with some evaporites progressively thickening northward. Besides this, there is an "inwards increase" which reaches a maximum thickness of 1.8 km along a broad meridional axial trough.
The Nubian beds of the Kufra Basin have a maximum thickness of 900 m and include mostly cross-bedded

Figure 14.1: Groundwater contours and flow directions in northeast Africa.

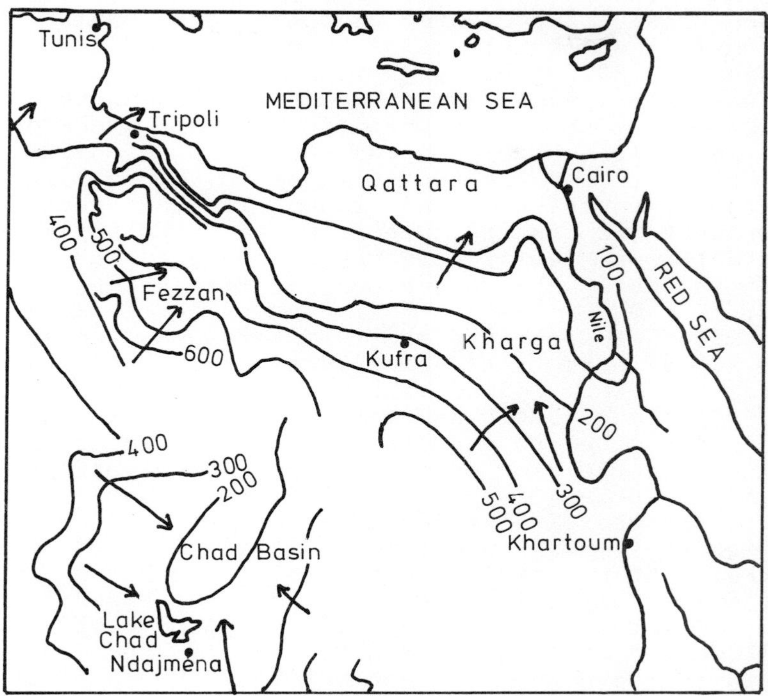

sandstones, subordinate shales and conglomerates which are fluvio-glacial and limnic origin. The Cretaceous sandstone aquifer of the Kufra Basin is phreatic with groundwater flow to the northeast discharging into the main Nubian artesian basin of Egypt. The water quality is fresh, with a TDS value of less than 100 mg/l. On the west and south, the basin is bounded by highlands which no doubt experience a considerable run-off. It is speculated that recharge from this run-off up to 20% of the total annual rainfall over the elevated areas may take place. This is roughly equivalent to the computed order of outflow from this basin of between 70 to 160 Mm³ every year. Based on transmissivity and piezometric gradient. The rainfall over the main basin is probably less than 3 mm per annum, so that vertical recharge is highly unlikely. In the north of the Kufra Basin, the age of the water exceeds 30,000 years. Hence, the phreatic aquifer cannot be in equilibrium with current recharge, although equilibrium may be approached in the marginal recharge areas.

In the south of the Sirte Basin, the groundwater is also fresh, but becomes brackish to saline in the north in the flow direction. Known groundwater ages range from 30,000 to 5,000 years B.P. and fall into three principal groups. Recent recharge water may exist in the uppermost phreatic levels. Recharge to the aquifer system in the basin is effected by lateral inflow from the Kufra Basin, marginal recharge through highland run-off to the west and southwest and direct input from precipitation in the north (Figure 14.1). Discharge occurs in the north into sabkhas between the Gulf of Sirte and Qattara (6).

In order to irrigate the Mediterranean coast of Libya, a huge irrigation network is being constructed. This follows the discovery in the middle 1970s of a gigantic aquifer in the southern desert. It is estimated to contain an amount of water equivalent to the flow of the Nile over 200 years. The plan envisages pumping 476,000 m^3 of water every day, transporting it north to water the rich, but desiccated soils along the Gulf of Sirte. Around Kufra, wheat, barley and alfalfa are being grown already. Initially, it is intended to expend some 3.3 billion US$ in order to convert over 1.19 million km^2 of land between Benghazi on the east and Sidra on the west into lush farmland. A second phase might extend this further to Tobruk on the east and Tripoli on the west, thus including almost all of coastal Libya. At first, 1,880 km of pipeline will be used and enormous quantities of concrete will be needed, this having inspired some to call it a "concrete Nile". At present construction is proceeding on the Tripoli-Benghazi section and on parallel lines from the coast at Brega south to the Sarir Well Field together with a single line to the Tazerbo Well Field. Later construction will carry a separate branch to Kufra and another line is proposed from Tripoli south to the Fezzan Well Field (7).

In the Sarir Calanscio Plain, in addition to the the well field mentioned above, there is an oil field. The region lies some 600 km southeast of Benghazi, has a moderate temperature and is highly arid. The southern part of the well field contains 157 wells, up to 300 m deep, which provide usually fresh water in quantities of 75 l/sec. The main groundwater body is considered to be 40,000 years old, but the age of the water table aquifer is probably not more than

5,000 to 2,000 years.

Figure 14.2: Hydrogeologic section in Egypt.

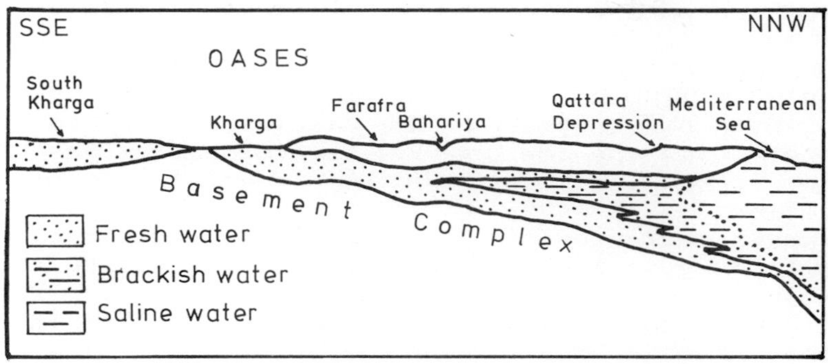

The Egyptian groundwater aquifers of the Farafra Oasis and neighboring much smaller Abu Monquar depression may belong to the huge multilayered Nubian artesian basin of monoclinal from which covers eastern Libya, north and northeastern parts of Sudan and Chad together with the whole of Egypt to make up one of the largest such basins in the world. The groundwater recharges comes from Chad (Figure 14.2). In this hyperarid region, large areas have an estimated mean annual precipitation of 1 mm, so that only in the southwestern mountain areas, to the south and perhaps on the eastern border of the Nubian Basin can there be some groundwater recharge form this source. During earlier humid conditions, rainfall recharge must have been considerable. No doubt, artesian or pumped groundwater in existing oases derived from a slowly discharging system filled in Holocene times. If today no recharge occurs, then the groundwater gradient must be attributed to other causes. Another possibility is that the groundwater is recharged by an assumed regional flow from southern marginal areas. The matter can be resolved by calculations involving former and recent flow conditions in areas where groundwater recharge is and was possible. Consequently, a numerical groundwater model for the Nubian aquifer system in the eastern Sahara was constructed to simulate long term regional groundwater flow. As regards this, incomplete information of the natural processes affecting this is the major handicap to its validity. In addition, transferring available data may entail mistakes. This model is one-layered, which, since upward leakage occurs in the confined northern part, appears to be unsatisfactory since a two-layered approach, at least, is thus required. It has also proved extremely difficult correctly to define the necessary forcing functions. In earlier models, a steady-state flow situation was assumed, but this is unjustifiable. In attempting to cover the entire Nubian aquifer system, the vast area involved together with the low groundwater velocities may well make it impossible to obtain appropriate forcing functions.

Regarding groundwater discharge, precipitational input today is probably at its minimum, taking place mostly in the Ennedi and Tibesti Mountains. It is believed that, although the annual rainfall is far below that of Western Europe, its intensity may facilitate some groundwater recharge. Recharge occurs from the Nile, as has been shown in the Gezira area near Khartoum. At Dakka, groundwater has been recorded as discharging into the river. During the pluvial periods, the recharge from precipitation must have been substantial. Apropos evaporation, discharge through this mechanism can only take place when the water table is near the ground surface, say 20 to 50 m deep.

There was probably a lake at Kufra 3,000 years or so ago with its surface about 5 m above the existing groundwater level. Similar larger lakes existed at Kharga. This indicates that the water table has declined rapidly in the area (8).

Nevertheless, by the time of construction of the Ibis Temple during the occupation of Egypt by the Persian King Darius during the period 531 to 436 B.C., the groundwater table had risen appreciably. There was decline thereafter and the local people obtained water from infiltration galleries and wells which reached 300 m in depth. In 1954, ten deep wells were drilled to more than 650 m in Kharga. One, at Baris, hit the basement and produced no less than 12,000 m³ daily, at least initially. The water is artesian and has a temperature of 36° to 38° C. Its output increases as a function of depth. The total concentration of dissolved solids is low, at 100 to 500 mg per liter approximating that of the Nile. Interestingly, the oxygen-18 values of Kharga well waters, which range from -10.3 to -11.5‰, also approximate that of the Nile which is -8.8‰. There is also a high nitrogen gas content. The excess of this is attributable to the Würm glaciation when a sea-level drop was accompanied by contribution of groundwater to the Neonile. The later climatic amelioration entailed a reversal, the river contributing to the aquifers and carrying air into them. Of this, the oxygen went to oxidize trapped organic material and hence a nitrogen excess remained.

Salt marshes and sabkhas have been traced back to such former lakes, these contemporary features being located over the confined part of the Nubian aquifer system, hence lacking direct connections with it. Nevertheless, a definite influence on the regional groundwater gradient is possible.

It is possible to infer a discharge from older sandstone sequences to younger aquifers of the Sirte Basin at a geological sill north of Kufra and the former estimated its magnitude of discharge to be appreciable. With respect to the Farafra Oasis, this is structurally a doubly plunging anticline with a pair of Late Cretaceous water-bearing beds. There is the Campanian-Maastrichtian chalk and compact limestone and the underlying Nubian strata. The densely jointed chalk is widely exposed in the northern area, has a maximum thickness of 175 m and grades into a marly facies to the south. Its groundwater is mostly fresh, but some wells have high salinity. The compact limestone underlies the chalk and reaches a maximum thickness of 50 m centrally. Cavities in it may be as much as 5 m across and the groundwater is fresh. In the Nubian beds, there are three water-bearing horizons, separated by shale aquicludes and the groundwater is usually fresh with mineralization increasing with depth until, at 300 to 600 m, it reaches 200 mg/l.

The Fayum covers 1,700 km^2, its almost flat surface sloping slightly northward finally to reach a level 55 m below sea. The surface is covered by a dense network of Nile-derived irrigation canals and drains into Lake Qarun. Lacustrine deposits blanket it. Many natural springs are found. To the southwest of it occurs the Wadi Rayan Depression which covers about 1,300 km^2 and has a surface strewn with sand dunes. The minimum depth of this is 59 m below sea level.

Groundwater occurs as perched aquifers in the unsaturated zone, as free surface groundwater within the lacustrine deposits at the Fayum and as artesian water recharging mineral and sulfur springs in Wadi Rayan.

The origin of these various types must be considered. At the Fayum, the springs are mainly recharged seepage from surface water in irrigation canals and drains through the formation of perched water tables at shallow depths. The groundwater of the lacustrine deposits is the result of lateral seepage from the Nile aquifer extending in the area to the east of the Nile-Fayum divide as well as by infiltration of surface water in various canals and drains together with downward percolation of some perched ground water. In the fissured Eocene limestone, the recharge comes from different sources, the area between the Fayum and Wadi Rayan being recharged mainly by seepage from the Fayum Depression. The Wadi Rayan mineral and sulfur springs are recharged by deep seated groundwater aquifers and probably by subsurface sandstones.

Almost all the groundwater in the Fayum is of meteoric origin. The increasing mineralization may result from dissolution and leaching of water-bearing rocks, cation exchange between water and rocks or concentration of salts.

As regards these groundwaters, samples from all over Egypt indicate that a high temperature geothermal resource does not exist there. This was determined by application of the silica, Na-K-Ca and Na-K-Ca-Mg geothermometers. Notwithstanding, thermal water samples were recorded from twenty wells (over 35° C) and four springs (more than 30° C), the hottest springs being found along the east coast of the Gulf of Suez at Ayun Musa (48° C) and Ain Hammam Faraoun (70° C). At the former, the oxygen-18 value of -7.6‰ is closely similar to that of the Nile at -8.8‰. This is unexpected in view of the high temperature of the spring waters. Ayun Musa is reputedly the site, Meribah, where the miracle of water from the rock occurred. According to Numbers 20 of the Pentateuch, the Children of Israel complained of the lack of water in the desert and " the Lord spake unto Moses, saying, Take the rod and gather thou the assembly together, thou, and Aaron thy brother, and speak ye unto the rock before their eyes, and it shall give forth his water, and thou shalt bring forth to them water out of the rock: So thou shalt give the congregation and the beasts drink". Moses obeyed, "smote the rock twice: and the water came out abundantly" (9).

Ayun Musa and Ain Hammam Faraoun are the optimal places for geothermal development. The Eastern Desert of Egypt, especially near the Red Sea, has above normal heat flow, therefore some geothermal potential as well. There is only one thermal well, namely Umm Kharga close to the Red Sea. This has a temperature of 35.8° C. In the major oases in the

Western Desert, namely Kharga, Dakhla, Farafra and Bahariya, the regional temperature gradient is under 20° C/km, but many wells tap deep artesian aquifers to produce vast volumes of water in the 35° to 43° C range and constitute a low temperature geothermal resource. No samples from the north of Egypt, including reported "hot springs", can be regarded as thermal.

A comparison between the hydrogeological situations of the Eastern Sahara and the Great Artesian Basin of Australia suggests the idea of autochthonous, thus fossil, groundwater in the Nubian aquifer system. Natural discharge of unconfined and confined groundwater into Saharan depressions causes an exponential groundwater decay during dry climatic periods and this diminution can be simulated by means of a non-steady hydraulic model which fits the groundwater ages obtained by isotopic dating.

Despite the occurrence of faults, the Nubian aquifers, extending over more than 2×10^6 km^2, is considered as hydraulically interconnected throughout and the groundwater body is stratified by interbedded clays and marls into an upper, unconfined and one or more lower, confined water units. The hydraulic gradient of some 0.5×10^{-3} is thought to induce groundwater flow from mountainous terrain in northern Sudan and northeastern Chad towards the Mediterranean. If this is true, the flow velocity would be roughly 6 m annually, a number based on the estimated mean horizontal permeability of the sandstones and a total porosity value of 0.1. This is contrary to the alternative idea that regional groundwater movement does not occur. In fact, stable isotope measurements show a significant difference in the heavy stable isotope contents between, on the one hand Kharga and Dakhla Oases and, on the other, Bahariya and Farafra Oases. The former are lighter than the latter to the extent of -3 ‰ or more in deuterium and 0.5 ‰ in oxygen-18. These differences may reflect separate water bodies and perhaps a related groundwater movement (10).

If the groundwater of the Nubian aquifer system originates only from inflow on the southern rim of the eastern Sahara and is allochthonous, then its age should continuously increase by 150,000 years or more along the flow distance of 1,000 km from south to north, assuming a velocity not exceeding 6 m annually. Because waters older than 30,000 years or thereabouts cannot be dated using radiocarbon, the radioisotope chlorine-36 may be applicable and has been utilized in Australia. This isotope is produced in exposed rocks by the spallation of potassium and calcium and the neutron activation of chlorine, being released locally in significant amounts by weathering. It seems that the level of chlorine-36 which accumulates in groundwaters by subsurface neutron activation only if their age exceeds one million years. This may well be the case in some of the aquifers because an 850,000 year He-Ar date has already been recorded. If so, the much longer half-life of chlorine-36 at 301,000 years could extend the dating range back by several orders of magnitude.

Sudan, the largest country in Africa and comparable in size with Europe, is totally dependent on groundwater for the activities of pasturalists and agriculturalists over vast areas of arid and semi-arid land to be made possible. In the north, rain is extremely infrequent and desert conditions prevail, so that, away from the river, wells are the only available source of water. In the south, there is a tropical continental type climate and, in the extreme south, this turns equatorial. Both these zones have a dry season which becomes more intense and longer northward. From October to June, the humidity declines over much of the country to reach a low of 10 % in the northern Sudan. There is high insolation and therefore high ground temperature resulting in very high evaporation rates of the order of 10 mm daily in the dry season of Central Sudan.

Water conservation is effected in many places by using open earth-dammed reservoirs (haffirs), but their efficiency is far from high because of evaporation and the erratic distribution of rainfall. Consequently, drinking water is often transported on camels, donkeys or tanker trucks, thus being expensive to buy. Along the Red Sea, the shell fishermen of Dungunab are believed to spend almost a third of their income on it, after which they blend it with brackish well water. This underlines the significance of groundwater in Sudanese life. The amount withdrawn from the various aquifers is difficult to estimate, but almost 3 million m^3 are taken yearly from the "Nubian" and Umm Ruwaba Formations in Kordofan alone.

The major water-bearing formations in Sudan in order of importance are the Nubian sandstones, the Umm Ruwaba Formation, the Gezira Formation, Pleistocene and Recent unconsolidated deposits, sedimentary formations of the Red Sea Littoral and the Basement Complex. The Nubian sandstones comprise bedded, flat or gently dipping conglomerates, grits, sandstones and mudstones. Groundwater usually occurs in the pebble conglomerates. The thickness of the sandstones rarely exceeds 450 m and it is by far the best aquifer in the Sudan, the water table lying between 90 and 120 m deep, except near rivers. There is recharge from these, including the Nile, as well as from rainfall which becomes more important southward.

The Umm Ruwaba Formation consists of unconsolidated sands, clayey sands and clays of fluviatile or lacustrine origin in a poorly sorted state and more than 300 m thick in places, the whole resting unconformably on the basement or on other rocks. The water table is again found between 90 and 120 m in depth and receives recharge by rainwater, seepage of water from wadis and the White Nile. The water is sometimes saline.

South of Khartoum, between the Blue Nile and the White Nile, the Gezira Formation includes clay beds overlying highly porous and permeable sands and gravels in which the groundwater occurs in lens-shaped, interconnected bodies of the latter. The thickness of the saturated strata is between 20 m and 25 m and the depth to the top of this zone is variable from 7 m near the Blue Nile to 52 m in the

Figure 14.3: A bird`s eye view of braided wadis in the Eastern Desert north of the Galalas, Egypt.

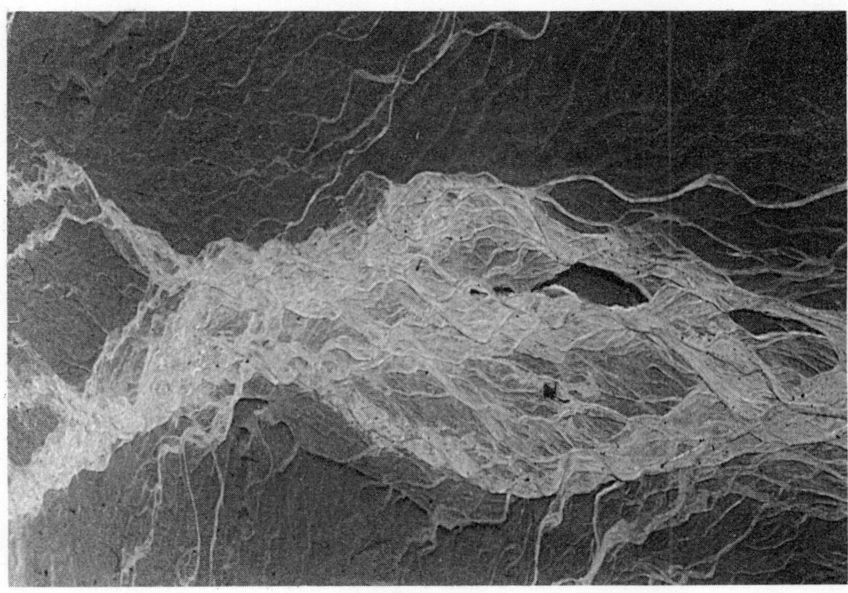

center of the Gezira. A pumping test at S.G.S. B.H. 1751 Hasaheisa showed that this formation is capable of producing about 1 m^3/sec. While the Gezira Formation is a good aquifer, it is restricted in area.

The Pleistocene and Recent material includes loose gravels, sands, sandy clays, silts and clays which accumulated in many different environments such as river terraces, deltas, wadis, etc. Obviously, their water-bearing capacity is extremely variable (Figure 14.3).

The sedimentary formations of the Red Sea Littoral include salt and anhydrite beds which are folded and faulted and often covered by superficial deposits, especially around Port Sudan where there is the greatest demand for water. These characteristics make exploration difficult and very little is known about the water potential. In addition, overpumping near the coastline causes salinity steadily to increase through salt-water intrusion. The basement complex produces very little water except where deeply weathered, jointed or fractured. Run-off may collect in large fissures and water be stored for usage in the dry season.

In northwestern Sudan, at Gebel Uweinat, pockets of water are sometimes found on the surrounding plain at Ain Dua and derive from occasional rainfall which seeps downward through fractures. Spring water is emitted at the adjacent Karkur Murr, probably along an unconformity in sandstones, later accumulating in gravels and sands in the wadi floor. There are numerous water holes as well, e.g. at Gebel Arkenu, Erdi and Ennedi. To the south is Wadi Howar, excavated near its headwaters in crystalline rock and downstream in the Nubian sandstones. During the rains, pools of water, such as Lake Tundur, have been recorded in its upper part and, in spate, vast quantities of sand and gravel are transported so there is a well-developed valley fill. It has been calculated that 1.5 km^3 of water pass every year into the superficial loose deposits on the floor of the wadi. It has been suggested that the water in the southern part of the Libyan Desert is derived from this and similar sources, but the disposition of the basement and younger rocks makes this highly improbable. Most likely, the bulk of the water in the Libyan and Western Deserts is fossil, of Plio-Pleistocene and earlier age. To the north of the Wadi Howar, in sand-covered areas there are small mud pans with patches of Acacia which indicate near-surface water, this mud reducing evaporation and conserving moisture in underlying porous rocks. Between Wadi Howar and the Nile, there is a number of oases and wells such as Selima Oasis, the water holes of Laqiya el Arbain which relate to anticlinal flexures in the Nubian sandstones, Bir Natrun, Bir

Figure 14.4: Contact springs flowing from Nubian sandstones on impervious basement and supplying a perennial lake in the volcanic Malha Crater, Sudan.

Sultan and others. The movements of groundwater in the lava plateaus of Meidob and Berti are governed by interbedding of porous and less porous flows. Water occurring in calderas and craters seems to be derived partly from rainfall, partly from ground-water and partly from magmatic sources. On the west side of Meidob, the Malha Crater has a salt lake in it and on its sides, there are springs emerging at the contact between the Nubian sandstones and the Basement Complex. The level of this lake is said to change suddenly, this sometimes being accompanied by strange noises from below (Figure 14.4).

In the south, around Kordofan and adjacent Darfur on an area covering 375,000 km² of Sudanese savanna, may be found Phanerozoic beds underlain by Pre-cambrian basement. The former include the Nawa Formation which is possibly Paleozoic and certainly Mesozoic, the Late Cretaceous Nubian sandstones, the

Cenozoic Umm Ruwaba Formation and loose superficial deposits. Few wells have been drilled in the sparsely populated area. Nevertheless, 2.7 billion m³ of water were extracted from the five discrete aquifers of the Nubian and Umm Ruwaba beds in 1962. Of course, there are many thousands of hand-dug wells which often run dry after the wet seasons.

In the Gezira, the main irrigated area of the country, potable groundwater is vital for drinking because many canals are infected with snails carrying Bilharzia as well as being insanitary. Most of this is obtained from the main aquifer which is the Gezira Formation.

The Gash River starts as the Mareb near Asmara in Eritrea, floods annually and transports huge quantities of gravel, sand and silt which deposit to form its delta in Kassela Province. This area is noted for irrigated fruit gardens and cotton is also grown. The aquifers in the highly variable delta deposits seem to be interconnected, the optimum beds being sands and gravels with saturated thicknesses of up to 30 m and the depth to water being from 6 to 18 m. The groundwater movement is toward the northwest of the delta, the gradient of the static water surface being 1.9 m/km. The main recharge is by seepage and the quantity of water stored in the 100 km² delta is 375 km³.

In the coastal strip near Port Sudan, the Khor Arbaat has a catchment area of 4,000 km², the rainfall being irregular and low, usually less than 100 mm annually. The mean yearly imput is believed to be 10 million m³, the quantity of water stored in the aquifer being 48 million m³ and the groundwater table lying some 5 m below the surface. Parts of the area have a perennial surface flow where there are narrow valleys and the alluvial fill is too thin to remove discharge by underflow.

Groundwater from Khor Baraka leaves the Red Sea Hills through a fanglomerate in the Tokar Delta, reaching the Red Sea with decreasing slope. Every year from July to September, a number of floods of low mineralized water from the Ethiopian mountains passes the wadi and contributes to the recharge of the aquifer. The arid Sudanese climate has a high evaporation potential exceeding 4,000 mm a year and the area suffers a very low precipitation of less than 100 mm annually. The groundwater becomes increasingly mineralized by evaporation directly through the capillarity of the fine-grained cover. The process is accelerated by saline intrusions from tributary wadis in the coastal area during the very low precipitation period from October to February. Apart from this increasing downstream salination, the groundwater chemistry alters by ion exchange in the aquifer. Ion exchanging minerals of the volcanic components of the bedrock and its talus sometimes change the nature of the groundwater to produce sodium bicarbonate and sulfate ions in it. Also, concretions and a calcareous hardpan develop in the aquifer. This classification of aquifer waters is very important in defining the thoroughly flushed fresh water zone as opposed to the more mineralized, deeper aquifer and hazardous saline water intrusions from tributary khors in connection with water supply to the coastal area (11).

WATERS OF THE LEVANT

The water resources of the Sinai Peninsula derive from rainfall, which occurs mostly in the north during winter when most of the 300 mm annual precipitation takes place, springs, of which only two are important. One is Ain El Gedeirat which emits 1,000 m^3 daily. The other is Ain Qadis yielding some 90 m^3 over the same period of time and also deriving from the same aquifer. Groundwater may be obtainable from the highly fissured basement complex, but is pumped from depths of 875 m from drilled wells in sandstones at El Hassana, El Hodira and Abu Darag as well as at depths between 330 m and 490 m at Ayun Mousa. The Mesozoic and Tertiary limestones also yield fresh water, for instance the Nakhl wells produce it from the basal beds of the Early Eocene chalk which is jointed and fissured. The Quaternary coastal sandstone comprises cavernous calcarous layers containing water under semi-artesian conditions, this having a highish salinity of up to 4,000 ppm. The deposits on the alluvial plains fringing the Gulfs of Suez and Aqaba contain water which is fresher toward the east, salinity increasing with pumping. Finally, the coastal sand dunes comprise a limited Holocene reservoir along the Mediterranean, this aquifer being supplied by winter rains. It is arranged in a series of fresh water lenses floating on seawater (12).
Groundwater potentials of the Arabian Peninsula are restricted due to an arid catchment area and the regional dip of the Phanerozoic sediments to the northeast. The general situation can be seen from Figure 14.5.

Figure 14.5: Schematic hydrogeologic cross-section of the eastern Arabian Basin.

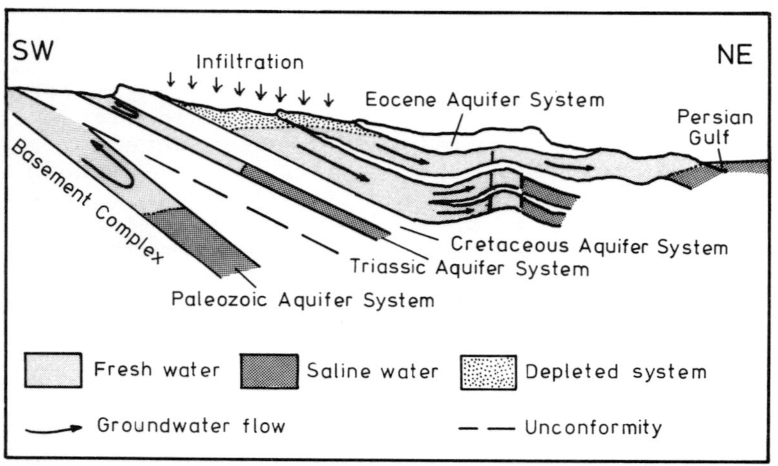

Zamzam is a groundwater well inside the Great Mosque (Masjeed Al Haram) at Mecca 20.6 m east of the Kabah. This source was discovered by Hagar, wife of Abraham, and her son Ishmael. The well is significant in Islam because drinking from and washing with its water are among its recommended rituals. Later neglect dried up the well which was rehabilitated by the Prophet Mohammed. Long after his time, probably following severe drought, the well dried up again, possibly 800 years ago, but was again reclaimed. A golden wall was built around it by the Ottoman Suleiman the Magnificent and later housed in a marble and lead dome. An iron mesh was placed 1 m below the water level to prevent people committing suicide. Since then, the well has been in continuous use (13).

The mosque is located in the Wadi Ibrahim and the catchment comprises Precambrian igneous and metamorphic rocks with alluvial cover 20 m thick. The aquifer is in fractured bedrock and the well, 30 m deep, is fed from three sets of such fractures in Precambrian diorite, these extending outside the surface water catchment. The vicinity of Zamzam is highly disturbed so that there is an efficient hydraulic connection. The dynamic level declines several meters during the Haj season, but recovery takes only half an hour after a drawdown of 10 m.

In Israel, the Judea Group aquifer may be connected with the underlying sandstone aquifer, indeed basically being recharged by it. Most of the water in these are non-renewable and the observed drop in water table level is due to drainage of the resource. Recently it has been suggested that the two aquifers recharge each other. There has been a considerable drop in the Dead Sea level of almost 100 m from 1930 to date. Changes in the static groundwater levels undoubtedly occur through exploitation, but there is a decline in the base level elevation (Dead Sea level) as well. In fact, if there is no intervention, the Dead Sea will decline at least another 50 m within the next half century which would induce a corresponding drop in the static water table and an average drop of 25 m in the entire catchment area. This would reduce its depth which now exceeds 300 m.

Apropos the stable isotope composition of the Dead Sea waters, these are moderately enriched in oxygen-18. The actual amount embodies a balance between the dilution through freshwater influx and the isotopic fractionation accompanying evaporative water loss and vapor exchange with atmospheric moisture. The values found for oxygen-18 range from +1.9‰ to +5‰, but the deuterium content lies close to zero. The former measure the isotopic activity ratio and the latter the concentration ratio.

Measurements were made on deeper water samples from the lake below 100 m in the late 1950s and they revealed a homogeneous oxygen isotope composition. Then, the Dead Sea was stratified with a large salinity gradient between 40 and 80 m deep. This separates a less dense water layer from a deep "fossil" water mass comprising highly saline water. Subsequently, the stratification weakened and disappeared completely by 1975. Generally, salinities and heavy stable isotope values move in tandem, but the correlation is not perfect. This is partly due to the fact that the diluting freshwater input comes from different sources with different degrees of depletion. One is the Jordan River and the oxygen isotopic composition of its inflow changed from -2.5 to -4.7‰ as the volume of this dropped from 1,135 to 350 km^3 annually. This was a result of the reduced contribution of the Sea of Galilee, also known as Lake Tiberias or Lake Kinneret, to the Jordan since their exploitation through the Israelian National Water grid. The important point is that the isotopic composition of the total inflow has altered by more than -2‰ for this heavy stable isotope. This is not simply because of the shift of isotopic composition in the actual river, but also because of the lesser weighting of the Jordan River water in the total water budget. It is recorded that the salinity of the surface waters of the Dead Sea increased from 250 g/kg^{-1} in 1960 to 276 g/kg^{-1} in 1979.

Saline springs and seepages discharging near the shore may go directly into the Dead Sea and represent an important salt influx. Their oxygen isotopic composition ranges from that of paleowaters, -7.5‰, to enriched ones of +2.1‰. However, such waters have practically no effect on the water budget of the Dead Sea and neither does the direct precipitation which is 75 mm on average annually. Although there is no available information, the isotopic composition of the local rain is assumed to be -4.5‰, the typical value for precipitation in the northern Jordan Valley at Degania.

The steady state isotopic composition of a similarly sized freshwater lake exposed to the same environmental conditions would accumulate the heavy stable isotope of oxygen to a value of +8.5‰. From this an estimate can be made of the isotopic value which must have prevailed in Lake Lisan, the precursor of the Dead Sea. If this lake was about half as saline as the Dead Sea, its anticipated isotopic composition would lie between +6.5 and +7.5‰, depending on the probable variations in the ambient parameters. This explains the increased enrichment in Lake Lisan carbonates relative to those precipitated in the existing Dead Sea. Antecedent to Lake Lisan was a freshwater Samra Lake which covered the northern portion of the rift, in all probability since the dawn of the Late Pleistocene (14,15).

Syria is part of an area with winter rains only, the mountains receiving precipitation up to 1,500 mm and central and eastern parts of the country receiving under 200 mm annually. Groundwater basins in Central Syria are independent and controlled by evaporation as is shown by the continuous increase in total dissolved solids in their centers. In much of northeastern Syria, the groundwater is locally strongly mineralized through saline solution by the aquifers despite regionally favorable drainage conditions. In this and other areas, there is a network of collection systems determined by tectonics. The dissolved solids result from evaporation, solution of rock and biogene contamination. Different temperature zones indicate major flows of groundwater and in basaltic regions of the Hauran, the groundwater is influenced by secondary jointing due to deep-seated structures. Primary jointing occurs due to the contraction of the basalts, but is not hydrogeologically significant because clays render it impervious. Between the Euphrates and Orontes, there are the closed basins of Aleppo and El Bab. Two groundwater aquifers are found in marly chalky rocks and in the limestone area of the western Aleppo Basin there are several aquifers traversed by joint systems. They contain confined groundwater and drain towards the western rift valleys.

The Ed Daou Basin is a large synclinal depression distinguished centrally by an extensive Quaternary plain 1,200 km² in area. Marine deposits range in age from the Middle Cretaceous to the Late Eocene, subsequently being folded during the Alpine Orogeny when a thick continental series was laid down in the basin. The catchment area covers 6,000 km² and has a semi-arid climate with a total average annual precipitation of 100 mm. Annual infiltration appears to be 60 million m³ of which most enters Quaternary sediments.

The basin in question lies just west of Palmyra, the magnificent capital city of Queen Zenobia (267-272 A.D.). An analysis of Palmyra thermal water from Ain el Qenah deriving from the Cenomanian limestone and yielding 40 m³ per hour showed that its temperature is 28.2° C and it contains 2,100 ppm of soluble salts, mainly chlorides and sulfates.

RADAR RIVERS

The "Bahr-bela-ma", the so-called large river without water, and the legendary "Zerzura", one of the presumed lost oases, are recorded in old stories dating back to ancient Egypt and stimulated European expeditions into remote parts of this country. None of these found such a feature, but the SIR-A did by the application of radar backscattering.

Equipped with the shuttle imaging radar (SIR-A), the Columbia flew over the greatest area of hyperarid terrain on Earth during its second test flight in November 1981. This is the Darb el Arbain Desert, where there are almost no superficial traces of fluvial action, eolian erosion and deposition being predominant. It is uninhabited except at some major oases. Its desiccation is due to its distance from the moist summer winds entering North Africa from the South Atlantic and also to its location far to the south of the storm tracks crossing the Mediterranean in winter. Rainfall anywhere occurs at about 30 to 50 year intervals. The 200 mm precipitation isohyet, which marks the northern edge of the tropical savanna zone in North Africa, is about 800 km south of the Central Arbain, but was not always so in the past. The SIR-A track enters this desert at the Chad-Sudan border, traverses the region north of the Merga oasis and south of the Gilf Kebir Plateau and Gebel Uweinat, goes over Egypt south of Bir Sahara and leaves it northeast of Bir Kiseiba.

The Arbain Desert is floored mainly by Cretaceous Nubia Formation and by low sporadic outcrops of granite and granite-gneiss of the Precambrian African Shield. The former is composed of cross-bedded sandstones and shales of which the clastics are informally called Nubian sandstones. Typically, the outcrops are wind-pitted, grooved and fluted to such an extent that they resemble rocks seen by the Viking Landers on Mars. There is practically no surface evidence for former major integrated drainage systems and vegetation is lacking, except in minor oases or wells termed "birs". The desert is mostly covered by yellow to slightly red windblown sand in vast, thin flat to undulating sheets with occasional enormous simple and complex barchanoid dunes which may be tens of kilometers long. Inselbergs without pediments, low outcrops, rare lake bed deposits and patches of gravel ventifacts render the barren landscape less monotonous.

Quaternary climatic cycles are evident. The beginning of the Pleistocene was already arid and episodic human occupation is manifested by Acheulian implements. This was at least 100,000 and perhaps as much as 300,000 years ago. Later, several periods of Mousterian and Aterian occupation took place between 100,000 and 40,000 years B.P. Reoccupation by Late Paleolithic and Neolithic people occurred about 10,000 years in the past. These Quaternary episodes of human settlement coincided with the pluvials. Sometimes they produced a savanna-like environment and permitted many small playas and intermittent streams to exist. Around them, many habitation sites have been found. Domestic animals appeared and there was local agriculture. By about 5,000 B.P., hyperaridity set in again and was followed by abandonment of the Arbain Desert.

The SIR-A track is particularly relevant to the Selima Sand Sheet, one of the most barren places on Earth. It comprises unweathered, finely laminated eolian sand commonly under 10 cm and usually not more than a few meters thick. In places, the sand overlies yellowish to reddish, pebbly alluvium of which some of the soils have been dated from artifacts and ostrich egg shells as about 8,500 years old. SIR-A penetrated the Selima Sand Sheet as well as smaller dunes migrating across it and drift sand in the Arbain Desert to reveal a remarkably different and mainly fluvial subsurface terrain known before only to Stone Age peoples and maybe earlier hominids.

Large-scale, integrated stream patterns once emerged from the now practically dead wadis of the 1,200 km² and 300 m high Gilf Kebir Plateau on to the flat plain flooring the Arbain Desert. Some of these defunct river beds were reconstructed by tracing their apparent courses between remnant interfluves recognized on Landsat images. The Wadi Eight Bells stream network drained at least 3,400 km² on the pediplane south of the Gilf Kebir Plateau. The former course of this master stream was traced on Landsat images southward to approximately 200 km beyond the retreating scarp of the plateau and to within 200 km of the SIR-A track in the northwest Sudan.

Presumably the Gilf Kebir is a relict drainage divide dating from the beginning of the erosional removal of Tertiary, Mesozoic and Paleozoic strata from south-central Egypt which succeeded the retreat of Late Eocene seas from the eastern Sahara. Courses of now defunct streams, which flowed north from the Gilf and probably carried some of the sand now migrating southward in the Great Sand Sea, have been recognized both from Landsat observations and on the ground. This reinforces the idea that the Gilf Kebir was a major drainage divide in the Arbain Desert, also in the entire eastern Sahara of western Egypt, eastern Libya and northwestern Sudan. To the west, a similarly large, integrated fossil drainage system probably of Late Quaternary age and formerly emanating from the north flanks of the Tibesti Mountains has been identified. It has been traced northward for over 1,000 km toward the Gulf of Sirte.

The Wadi Howar marks the south boundary of the Arbain Desert. Major streams flowing from the Gilf Kebir highlands in Middle to Late Tertiary time may have reached it and thence discharged into the upper Nile. On the other hand, its streams may have reversed their courses at least once due to disturbances in the Gebel Marra. Then, they flowed into Lake Chad together with the discharge from the former Gilf Kebir catchment areas. They would also have contained contributions from defunct west and south flowing stream courses, the radar rivers discovered by SIR-A.

The present regional slope of the land in south-western Egypt is generally northward. Capture of some of the wadis of the Gilf Kebir Plateau by the south-flowing Eight Bells drainage and by other streams that eroded headward into the west side of the Gilf during less arid episodes suggests former steeper gradients to the south and west than to the north in these areas. If the Gilf Kebir Plateau were indeed once a major divide separating east and north- from south-flowing drainages in the Arbain Desert, the latter streams could have discharged into the Bodélé Depression in the Borkou region of northern Chad without ever having been linked with Wadi Howar. Such drainage directions would have had to predate uplifts of the Uweinat and Tibesti complexes and associated volcanism that must have disturbed the regional gradients in the eastern Sahara. The Columbia radar swath extended from the Mourdi Valley in northeastern Chad across northwestern Sudan to Kom Ombo near Aswan on the Nile. Six areas were mapped in some detail and chosen because they possess hitherto unknown or ambiguous geological features. The delineation of the latter is dependent on the ability of the SIR-A radar to penetrate the practically continuous eolian blanket veiling this hyperarid region. The most striking result of the investigation was that the SIR-A pictures, even with only preliminary processing, and the radar terrain maps constructed from them showed major subjacent features. These were undetectable even on specially processed Landsat pictures on the same scale. The major radar river systems may have flood plains as wide as or wider than that of the Nile Valley to the east. The broad trunk valleys have only a few equally wide tributary valleys which appear to be truncated, perhaps by long episodes of deflation. The stubby broad valleys and much smaller super-imposed networks of wadis have junction angles with the trunk streams which indicate west-, north- and southward drainage. These earlier directions of stream flow do not accord with these buried water-ways ever having been connected with the present drainage of the Egyptian Nile. Perhaps the large valleys are relics of Tertiary systems draining the eastern Sahara long before general aridity set in during the Early Quaternary and also long prior to the integration of the Nile.

Ancient Tertiary rivers which flowed in the direct-ions shown by the radar images are quite inconsist-ent with the present trend of the water table in Egypt. This slopes gradually from high ground in the southwest northward toward the Qattara Depression and the Nile Delta. This does not necessarily invalidate the argument, supported both by Landsat analyses and gound truth, for the earlier existence of larger scale south- and eastward-flowing drainage systems which emanated from the Gilf Kebir. In fact, the SIR-A evidence contributes to a corroboration of an earlier interpretation of the plateau as a major pre-Pleistocene divide in the eastern Sahara. The pediplane of the Arbain Desert was carved by large and very energetic streams during the interval between the retreat of the Eocene sea and the start of the Quaternary aridity.

The dominating streams on the radar images have partly to completely buried floodplains including narrow channels and underfit tributaries. They indicate that parts of the water courses were repeatedly used, probably by intermittent running water, during Quaternary pluvials. Dark radar response areas mark the traces of some smaller channels reminiscent of billabongs which characterize relict drainage of semi-arid to arid Central Australia. Thus, the SIR-A pictures expose a series of palimpsest drainages in the Arbain Desert which are probably connected with a number of fluvial episodes. Some may be as old as the Oligocene, others perhaps being as recent as Quaternary pluvials coinciding with human occupation (16).

Archeological discoveries in the Western Desert revealed three main periods of human occupation here before prehistoric time. The earliest Late Acheulian artifacts, possibly of Homo erectus, are initially associated with ancient river deposits. Later, as the climate deteriorated, they were buried in tufa mounds, these being spring sites which some quarter of a million years ago were at the base of retreating cliffs with artesian aquifers. The Middle Paleolithic occupation, which may have been by Homo sapiens neanderthalensis, seems to be linked with both alluvial deposits and the dunes. The latter had commenced to invade parts of the Western Desert since the Early Pleistocene. Neolithic artifacts of Homo sapiens sapiens in the Arbain Desert are generally restricted to playa and eolian deposits, except where they may have been lowered by deflation on to older alluvial surfaces. This demonstrates the increasing environmental severity in Egypt as the country moved nearer to its present hyperaridity.

The radar pictures illustrate a fluvially-dominated ancient topography, but Landsat imagery better reveals existing superficial geologic units, mostly distinct types of eolian deposits. False-color Landsat images show that the Selima sand sheet is not one blanket of sand over alluvium, but comprises successive layers of overlapping sand. Mostly, these are featureless drifts and wind-truncated remnants of massed barchanoid dunes. Overriding all these are trains of actively moving barchans representing the most recent wave of sand penetration probably deriving from the Great Sand Sea to the north.

REFERENCES

1. SHATA, A.A., 1983. Management problems of the major regional aquifers in North Africa and the Arabian Peninsula. Sydney, Int. Conf. Groundwater and Man, 263-271.

2. BURDON, D.J., 1980. Infiltration conditions of a major sandstone aquifer around Ghat, Libya. In: The Geology of Libya, 2, ed. SALEM, M.J., BUSREWIL, M.T., 595-611, Acad. Press, London, UK.

3. THORWEIHE, U., SCHNEIDER, M. and SONNTAG, C., 1984. New aspects of hydrogeology in southern Egypt. Berliner geowiss. Abh., A, 50, 209-216.

4. BOWEN, R., 1986. Aspects of Egyptian hydrogeology. Water International, 11, 64-70.

5. PALLAS, P., 1980. Water resources of the Socialist People`s Libyan Arab Jamahiriya. In: The Geology of Libya, 2, ed. SALEM, M.J., BUSREWIL, M.T., 539-594, Acad. Press. London, UK.

6. WRIGHT, E.P., BENFIELD, A.C., EDMOND, W.M. and KITCHING, R., 1982. Hydrogeology of the Kufra and Sirte Basins, Eastern Libya. J. Eng. Geol., 15, 83-103.

7. BOWEN, R., 1986. Groundwater. 427 pp., Elsevier Appl. Sci. Pubs, London, UK

8. HEINL, M., HOLLÄNDER, R., 1984. Some aspects of a new groundwater model for the Nubian aquifer system. Berliner geowiss. Abh., A, 50, 221-231.

9. SWANBORG, C.A., MORGAN, P. and BOULOS, F.K., 1983. Geothermal potential of Egypt. Tectono-

physics, 96, 77-94.

10. SONNTAG, C., 1984. Autochthoneous groundwater in the confined Nubian Sandstone aquifers. Berliner geowiss. Abh., A, 50, 217-220.

11. LANGSDORF, W., 1981. Mineralisierung des Grundwassers eines Wadis unter wuestenhaftem Klima (Khor Baraka/Sudan). Z. deutsch. geol. Ges., 132, 637-646.

12. HAMMAD, F.A., 1980. Geomorphological and hydrogeological aspects of Sinai Peninsula, ARE Ann. Geol. Surv. Egypt, 10, 807-817.

13. BASMACI, Y., ALWASH, M., 1981. Zamzam well.I.Intern. Congr. Hist. Turkish-Islamic Sci. Techn., ITUE, 127- 132.

14. KAFRI, U., 1981. Relationship between Dead Sea and groundwater levels in the western Dead Sea catchment area. Geol. Surv. Israel, Current Research, 2, 90-94.

15. GAT, J.R., 1983. The stable isotope composition of Dead Sea waters. Preprint, Rehovot Isotope Dept, Weizmann Inst. Sci.

16. McCAULEY, J.F., SCHABER, G.G., BREED, C.S., GROLIER, M.J., HAYNES, C.V., ISSAWI, B., ELACHI, C. and BLOM, R., 1982. Subsurface valleys and geoarcheology of the Eastern Sahara revealed by Shuttle Radar. Science, 218, 1004-1020.

Chapter V-15

LIBYAN DESERT SILICA GLASS

"There was a sea of glass like unto crystal". Revelation, 23.

ASTROBLEMES FROM OUTER SPACE

Before World War II, an expedition to southwestern Egypt reached the Great Sand Sea and its members described this geologic enigma as lying about in greenish-yellow lumps and glittering in the sun. In the middle of the 19th Century, the desert glass had been seen previously by Haji Hossein on the caravan trail from Kufra to Dakhla, but, of course, it was known and used in Dynastic, and Predynastic times as well as during the stone ages. The first mention of the "verre libyque" in the modern era of science was by the French physicist A.J. Fresnel early in the 19th Century.

Libyan Desert Silica Glass or LDSG has attracted much interest since it was first described. Its occurrence, physical characteristics and chemical composition have been discussed in order to clarify its origin. This involved examining its technical properties and its use in Paleolithic life together with its regional and stratigraphic distribution in the Great Sand Sea.

In ancient Egypt, glass was described as melted stones and isolated examples of artificial frit as beads are found from the Fifth Dynasty to the beginning of the New Kingdom when glass was introduced on a large scale as an intentionally produced material. The earliest datable pieces carry the name of Tuthmosis III. Much Egyptian glass is blue which may well indicate that it was not an Egyptian invention since cobalt compounds do not occur in that country. The old glass is a soda-lime-silicate which was usually opaque, although the craftsman tried to

imitate gemstones. It is extremely interesting that the people regarded such artificial ornaments as much inferior to those provided by the gods. Real stones for jewellery came from minerals collected and mined in the Eastern Desert or Sinai and the presiding goddess was Hathor.
It is an open question as to whether or not the desert glass was known to the Egyptians in Pharaonic times. Their word "thnt" was generally used in reference to faience or glass, but the identity of a precious stone recorded as "thnt m" is uncertain. This is partly because it is recorded rather rarely, but, where it does occur it is listed among gemstones or as a material for amulets. Since it relates to glaze or glazed ware and glass, it may well include some form of vitreous mineral. Among the candidates proposed are transparent calcite, obsidian, rock crystal, crystalline gypsum and silica glass. Iceland spar is much too rare to merit special attention and rock crystal as well as obsidian are probably described by the words "mnw hd" and "mnw km". The present name of the latter comes from the Roman "obsidianus", a faulty reading of "obsianus" in Pliny, who stated that it was discovered by a person called Obsius.

Figure 15.1: A wind-blasted cobble of desert glass on an interdune deflational surface in the Great Sand Sea.

Silica glass covers at least 3,000 km² of the Western Desert and some lumps of it are the size of a human head (Figure 15.1). In addition worked flakes are often found. The material "thnt" came from "thnw", this having been situated to the west or the northwest corner of the Delta which may have been a possible source of silica glass. There is no evidence of the color of the real "thnt", although most of it was probably green or greenish-yellow. The natural glass was probably used ornamentally by the old Egyptians since "thnt m" together with "ds" is the material for five amulets which include a scarab and a figure of Thoth. The word "ds" may well mean "flint", but could also mean "white marble", "dark marble" or "yellow alabaster". An amulet of "ds" is actually quartz.

Initially, a volcanic or meteoritic origin was proposed for desert glass. However, it soon became evident that the texture and chemical composition are incompatible with a relationship to volcanic glass. Additionally, there was no volcanicity in the area at the time of formation (1).

Extraterrestrial connections are still being sought. Could it be that the glass arose "where meteors shoot, clouds form, lightnings are loosened, stars come and go", to quote Robert Browning? Then, it would have differentiated in the heat of a so-far unknown type of meteorite and subsequently landed on Earth. However, neither its non-aerodynamic shape nor its composition of high silica and water contents support this hypothesis. Alternatively, it may form from melting quartz sands due to heat generated by meteoritic impact. An iron meteorite from the region, which fell less than a century ago, does not appear to have produced any glass. The almost identical chemical composition and homogeneous texture of Lybian Desert Silica Glass over a vast area seems to militate against any relationship with lechatelierites, although astroblemes were recorded from southern Cyrenaica. A 12 kg iron meteorite, classified as a nickel-poor ataxite, was found by a nomad in the Western Desert about 30 km to the west of Aswan. This is the earliest recorded from the Western Desert. There are only a few other meteorite falls recorded in Egypt during the present century. One is that of El-Nakhla El-Bahariya in the northwestern part of the Nile Delta in 1911. This is composed of about a dozen fragments of a calcium-rich stony achondrite. Another is from Qantara on the Suez Canal. It fell in 1916 and is a 1.4 kg chondrite of intermediate type.

Figure 15.2: Generalized geological map showing the "glass area" and the location of Libyan astroblemes.

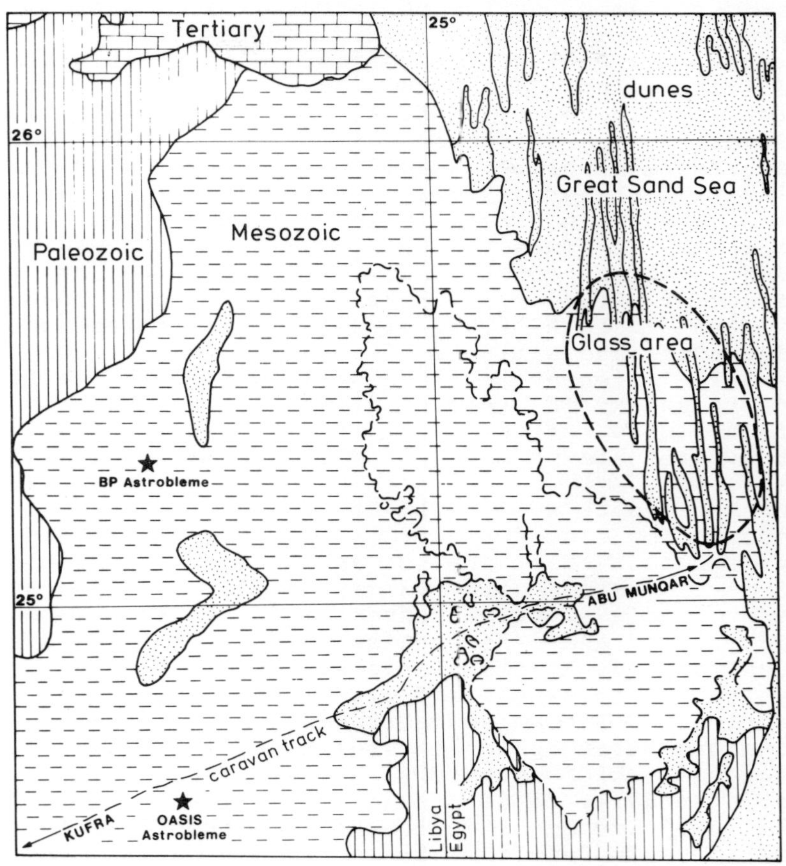

Black and dark grey impact silica glass in irregular fragments was found in great quantity outside the crater of Aouelloul, near Adrar in the western Sahara. This remarkable structure is 250 m in diameter with a rim rising up to 60 m above ground. The inclusions of sand in the glass suggest that the country rock has contributed largely to its composition and indeed may be the main ingredient of the Aouelloul glass. However, in its chemical composition and internal structure, such lechatelierites are completely different from desert glass, but rather similar to fulgurites.

A pair of large ring structures comprising low and concentric rock outcrops are situated in the Kufra Basin roughly 100 km west and 120 km southwest respectively of the so-called glass area. Originally identified by Landsat, they were first considered to be astroblemes or impact structures. Later, they were associated with the strewn field of the glass. The smaller "BP" structure consists of three circular rings about 20 m high and has a maximum width of 2.8 km. By contrast, disturbed rocks of the bigger "Oasis" structure reach a diameter of 11.5 km and embrace hills attaining more than 100 m in elevation (Figure 15.2). Despite the fact that both structures were carefully investigated, no meteoritic fragments, impact glass, nickel-iron spherules or even shatter cones and megascopic breccias were observed. However, quartz grains from sandstones exposed in both of the circular structures showed cleavage and planar fractures with prominent parallel orientations suggesting metamorphic effects due to shock. The resultant lamellae and the absence of definite traces of volcanicity within the ring structures favor meteoritic impact as the likeliest cause (2).

Further south near Gilf Kebir and Gebel Uweinat, several low concentric ridges of deformed sandstones are exposed. They are not as big as the astroblemes from the Kufra Basin and can usually be recognized as volcanic eruption centers. There are some which lack lavas and pyroclastics so that their origin is questionable. Another large elliptical structure 3.5 by 1.8 km in extent and with inwardly dipping beds was located 160 km eastnortheast of the Kufra Oasis. No shock metamorphic features were seen and the rock sequence, which is only slightly disturbed, may turn out eventually to be a syncline.

The coincidence of at least two astroblemes in the Kufra Basin and the strewn field of glass in the southern corner of the Great Sand Sea stimulated attempts to find a link between them. The glass only appears on sandy or gravelly desert floors and revealed some chemical and mineralogical evidence of derivation from the Mesozoic sandstones of the area. Hence, a connection with the astroblemes remains either indeterminate or merely speculative.

Silica glass from the Western Desert shows some resemblance to tektites, although it is found in larger pieces than them. It differs from the black Arabian silica glass in the Wabar Crater because this is definitely of impact origin, showing not only flowage, but also containing abundant spherules of nickel-iron.

Desert glass

A RIDDLE OF THE SANDS

Until recently, the desert glass was recorded solely from wind-scoured surfaces of the interdunal streets in the Great Sand Sea and therefore involved only sand-blasted, transported chunks or artifacts. Some idea as to how long such fragments have lain on the desert floor after exposure through erosion is given by their reactions to radiation. Cosmic rays interacted with these pieces and produced beryllium-10 and aluminum-26. Aluminum-26/beryllium-10 ratios gave a value of 3 or so. From this, it is concluded that these nuclides were the result of spallation within the glass. Considerations of cosmic ray flux, estimates of land surface elevation and beryllium-10 and aluminum-26 concentrations led to the inference that it has been exposed to such radiation for periods ranging from 200,000 to a million years (3).

The first hint as to the parent rock was derived by tracing glass clasts in a complex of sand and gravel alluvial deposits with varying degrees of red paleosol development. This was on the east side of a dune ridge near the southwest edge of the Great Sand Sea. This gravel bed covers quartzitic bed rock and is succeeded by lacustrine and eolian deposits. As an Acheulian hand axe was found on the top, it was assumed to be considerably older than 200,000 years. For comparison, fission track analysis showed that the glass ages reach 28.5 Ma. Only minimal concentrations of uranium, 1 ppm, and lead, 5 ppm, are present in the glass, thus necessitating the application of other dating methods. This is true also of K-Ar, from which a range of ages from 30 to 10 Ma were obtained. Consequently, the fission track approach appears to be most suitable. However, the results are far from satisfactory in view of the fact that the ages are usually too low and the decay constant extremely difficult to deduce (4,5).

There is no doubt that the clasts and chunks of glass occur in the basal gravel bed of the Cenozoic sequence. However, lacustrine deposits are superimposed and do not interfinger with pre-Neolithic whaleback dunes. Actually, they can be correlated with identical deposits on the western slope of the dunal ridge in a deflational street. Outcrops of gently folded Mesozoic quartzitic bedrock form irregular and rugged depressions near the southern end of this street. In such places, an abundance of flakes and artifacts of Neolithic to Paleolithic age can be found. Those not made from glass were produced from the dense, hard and outcropping quartzites. The uneven, rocky surface of the bedrock

Figure 15.3: Dunal ridge and wind-deflated street, exposing Neogene lake deposits superimposed on a gravel bed and Mesozoic bedrock.

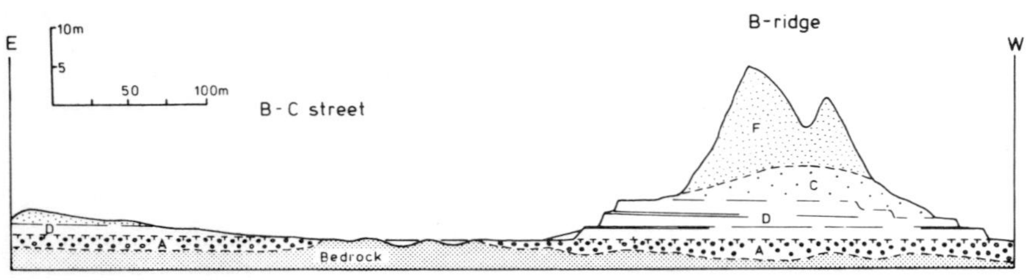

is covered by the aforementioned red gravel bed consisting of well-rounded pebbles of an average diameter of 0.5 cm in a red sandy matrix. All the pebbles are veneered with limonite and hematite and appear to be of rather local derivation. In the weathered debris of this horizon, 1 to 2 cm clasts and 5 to 20 cm chunks were observed together with pebbles of quartzitic sandstone and quartz. Most of the glass was more or less affected by wind corrasion. The red gravel bed is overlain by a lacustrine sequence about 5 m thick which comprises, brown, cross-bedded sands and thin, gray, partly silty layers. In some of the arenaceous beds, the depositional textures are nearly obliterated by a multitude of penetrating rhizocretions, whereas the clay layers are all dissected by desiccation cracks (6).

The sequence indicates the existence of ephemeral lakes which, during wet periods, even supported marginal vegetation. There is a great difference as regards grain size, sorting and texture between lacustrine and overlying eolian sands and whaleback and active dunes. This is despite the fact that large amounts of deflated material have been incorporated within the latter. The clays do not represent more than 15-20% of the lacustrine sediments. They include the usual mineral assemblage together with feldspars, hematite and some dolomite. Sporomorphs, diatoms and other plant fragments are also present (Figure 15.3).

In one street, the red gravel bed is widely distributed. Near the western dunal ridges, reworked glass is common and has been detected in situ. Unlike pebbles, it is not randomly distributed in the red gravel bed. Vitreous clusters nestle in partially sand-filled crevices which once were interconnected with the desiccation cracks in the mostly eroded lacustrine overburden. Unscoured by eolian sands, autochthonous glass is not as attractive in appearance as when found on the desert floor. Its rough, spongy surfaces replicate the sandy and pebbly walls of the crevices from which it is taken. There are loose contacts between the desert glass and adjacent sand grains or pebbles and no reactions at the interfaces are indicated. Thus, attached red soil can easily be brushed away, displaying a rough, dull surface.

The shape and in situ position of the material provide one answer to the riddle of its age. Desert glass is younger than the red bed, but is perhaps genetically connected with the overlying lacustrine deposits. It was formed in at least one unit by the precipitation of silica probably derived from groundwater percolation. With respect to its properties, the OH-concentration of 1,800 to 1,400 ppm is rather high and iron is mainly present in the trivalent state, which precludes any relationship with tektites or impact glasses. Another point against any connection between them is that they have very different dielectric constants, 4.2 for the desert glass and 6 to 7.4 for the tectites.

Figure 15.4: Major chemical components in volcanic glasses, silcretes, tektites, lechatelierites and industrial glass in comparison with desert glass.

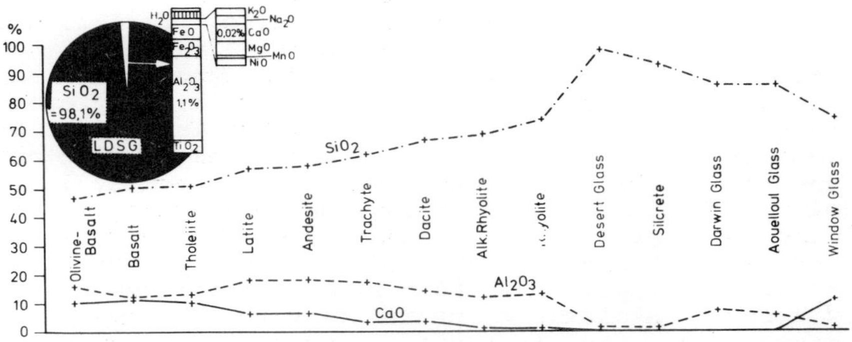

Desert glass is characterized by high silica content around 98%, easily measurable concentrations of alumina up to 1%, but only minute traces of the alkalis at 0.01 to 0.03 %, with a water content of approximately 0.1 %. It contains carbon, the concentration of which is higher in dark varieties at 0.01-0.02% than in greenish with 0.006-0.008%. As well as inclusions of carbon dioxide, brown streaks or patches can sometimes be seen in polished sections. Chemically, the desert glass is very similar to the sands and rocks where it occurs. Its viscosity around 1,000°C is much less than that of fused silica. This raises the problem of how the environmental sands could have been melted and fined if they really comprise the source of the glass. It is significant that partially fused material is not recorded from the site, hence such an origin is excluded. The same argument is applicable to other high temperature causes which have been proposed (7).

Organic compounds such as phytane and pristane have also been found in it. A zoogene source of the pyrolized organic compounds may be inferred from the rare occurrence of isomeres or the balanced proportion of unequal and equal numbered hydrocarbons. Up to a quarter of the hydrocarbons are saturated. Decomposition of organic compounds was prevented once they were trapped in the glass. Gas-chromatographic results from the lacustrine clays surmounting the red gravel bed indicated organic components similar to those in the glass, not surprising because both act as seals. The isotopic composition of organic carbon from desert glass is -24.3‰ for the carbon-13/carbon-12 ratio and indicates derivation from a C-3 plant association. Interestingly, tandem-accelerator mass-spectrometer analysis of the oxygen isotope ratio in the silica of the desert glass gave a value of +2.07‰, corresponding to a delta value of +32.4‰ on the SMOW scale. This falls within the upper range of ratios for sedimentary rocks and is quite different from that of extraterrestrial materials which is approximately -2 to +12‰. Clearly this rules out any meteoritic, lunar or other cosmic sources.

Dark or brownish desert glass contains palynomorphs. They are badly preserved and generally break into unidentifiable pieces when etched with hydrofluoric acid. Nevertheless, in thin flakes, sections or even Neolithic blades, they can be detected under the microscope. Such inclusions have not been recorded in the yellow-greenish varieties of it. There are spores and pollen indicative of the previous occurrence of ferns, grasses, pines and acacias.

Fungal and diatomaceous remains are found together with epidermal fragments of arthropods. These fossils rule out any possible extraterrestrial origin for the desert glass. Consequently, there can be no relation to the astroblemes of the Kufra Basin or any meteorite, hypothetical or otherwise. Rather, the occurrence in the sections and the components in the glass strongly suggest a direct relationship with a former lake, its sediments and the interrelated hydrochemical regime.

Similar conditions are presently found in Lake Magadi in a volcanic terrain in the Gregory Rift Valley of Kenya where the hydrous sodium silicates magadiite and kenyaite are found in an area far exceeding 100 km^2. These are well-crystallized layered silicates which convert to a crystalline hydrate of silica in dilute acids. Concretions in the magadiite bed, which is of Late Pleistocene to Recent age, comprise central kenyaite or chert surrounded by kenyaite. The bed itself is most likely a chemical precipitate from alkaline brines of the precursor of Lake Magadi and this mineral is converted to kenyaite and ultimately to chert by percolating waters. The Magadi chert is frequently found in dessication-produced polygons extending up to 50 m across as a result of periodic drying out of the ephemeral water body. It is in the cracks framing such features and, being more resistant, the polygonal-shaped rims may stand up as much as 2 m above the lake floor. This may be a mechanism of formation of bedded cherts by inorganic precipitation. Alternations between silica-rich and iron-rich bands of iron formations could be due to concentration cycles in alkaline lakes.

The precipitation of magadiite takes place in annual increments in a stratified lake at the brine-epilimnion interface, the maximum duration being estimated at 4,000 to 6,000 years. Now, most of the Magadi waters are supersaturated with regard to magadiite and kenyaite, although no precipitation has been observed. Chert chips occur in intraformational gravels and show that the conversion to chert was initiated during the magadiite precipitation interval (8).

The development of the Egyptian desert glass succeeded the deposition of sand and clay in a lacustrine basin under warm and arid climatic conditions in Neogene times. In its water, diatoms flourished and the shores were fringed by shrubs, brushes and herbs. Considerable amounts may have evaporated and periodically there must have been no lake at all. Anyhow, such an ephemeral situation would have severely affected floral and faunal communities, not least as a result of fluctuation in salinity. Once emerged, the clay beds of the former lake floor would have shrunk and cracked by desiccation.

Some brackish water might have escaped evaporation within the red gravel bed at the base of the lacustrine deposits. This groundwater would have had high pH values partly because of the ammonia derived from the anaerobic decomposition of organic matter. High pH values would have also occurred in percolating water and controlled the solubility of siliceous biogenes such as diatoms, sponge spicules and phytoliths within the lacustrine deposits as well as the ionization of orthosilicic acid. In the lower part of the aquifer, high concentrations of dissolved silica may have been maintained for rather long periods of time, particularly if nucleation of silica prevented supersaturation. Due to the growth of minute silica solids, colloidal solutions gradually accumulated in the aquifer.

In a succeeding humid phase, the lake basin would have been replenished with fresh water promoting both planktonic life and marginal vegetation. Biogenic metabolism and decomposition would have considerably increased the concentration of dissolved carbon dioxide in the lake water. Through desiccation cracks in the underlying lacustrine deposits, acidic lake water could have contacted alkaline groundwater. On account of the resulting electrolytic effects, the sols of silica would have attained a minimum conductivity and solubility, hence coagulated. Within the silica gels, hydrocarbons and microfossils might have been trapped as part of the suspended load. The semi-solid consistency during glass formation can be noted from its inclusions such as tear-shaped gas bubbles as well as from its unweathered sculpture, it being a true cast of the adjacent sediments.

Climatic changes and disrupted drainage systems brought an end to the formation of silica glass. Most of the lacustrine deposits suffered erosion, exposing the somewhat more resistent red gravel bed over a large area. This is the reason that allochthonous glass is so commonly found on the desert floor, in this way delineating the boundaries of a huge, desiccated Neogene lake basin (9).

Such a sedimentary origin is implied by the X-ray diffraction pattern showing its almost amorphous composition. However, there are always indistinct peaks of cristobalite and tridymite. In some samples, devitrification effects can be seen macroscopically through the ocurrence of white spherules of about 1.5 mm diameter. The transformation of the glass to a stable cryptocrystalline quartz aggregate is initiated by such intermediate phases. Perhaps the chalcedony nodules so often found in the red soils along the Darb el Arbain and covered by the Selima Sand Sheet are similar to it in origin, but older.

REFERENCES

1. GIEGENGACK, R., ISSAWI, B., 1975. Libyan Desert Silica Glass, a summary of the problem of its origin. Ann. Geol. Soc. Egypt, 5, 105-118.

2. UNDERWOOD, J.R., FISK, E.P., 1980. Meteorite impact structures, south-east Libya. In: The Geology of Libya, 3, ed. SALEM, M.J., BUSREWIL, M.T., 893-900, Acad. Press, London, UK.

3. GIEGENGACK, R., KLEIN, J., MIDDLETON, R., SHARMA, P., UNDERWOOD, J.R., jr. and WEEKS, R.A., 1985. Exposure ages of Libyan Desert Glass determined from in situ production of beryllium-10 and aluminum-26. 12th Int. Radiocarbon Conf. Trondheim., Abstr. 143.

4. GENTER, W., STORZER, O. and WAGNER, G.A., 1969. New fission track ages of tektites and related glasses. Geochim. Cosmochim. Acta, 33, 1075-1081.

5. ROE, D.A., OLSEN, J.W., UNDERWOOD jr., J.R. and GIEGENGACK, R.T., 1982. A handaxe of Libyan Desert Glass. Antiquity, 56, 88-92.

6. HAYNES, C.V., 1982. Great Sand Sea and Selima sand sheet, eastern Sahara: Geochronology of desertification. Science, 217, 629-633.

7. FRISCHAT, G.H., KLÖPFER, C., BEIER, W. and WEEKS, R.A., 1982. Glastechnische Untersuchungen an libyschem Wüstenglas. Glastechn. Ber., 55, 11, 228-234.

8. EUGSTER, H.P., 1969. Inorganic bedded cherts from the Magadi Area, Kenya. Contr. Mineral. and

Petrol., 22, 1-31.

9. JUX, U., 1983. Zusammensetzung und Ursprung von Wüstengläsern aus der Grossen Sandsee Aegyptens. Z. deutsch. geol. Ges., 134, 521-553.

L'ENVOI . . .

A Ramesedan caricature on a rock slab, perhaps the first record of a geologist at work, but more likely a quarryman or sculptor.

GEOGRAPHIC INDEX

"How index-learning turns no student pale,
Yet holds the eel of science by the tail".
Alexander Pope, 1688-1744.